SCIENCE & REASON

SCIENCE & REASON

Henry E. Kyburg, Jr.

New York   Oxford
OXFORD UNIVERSITY PRESS
1990

Oxford University Press

Oxford New York Toronto
Delhi Bombay Calcutta Madras Karachi
Petaling Jaya Singapore Hong Kong Tokyo
Nairobi Dar es Salaam Cape Town
Melbourne Auckland

and associated companies in
Berlin Ibadan

Copyright © 1990 by Henry E. Kyburg, Jr.

Published by Oxford University Press, Inc.,
200 Madison Avenue, New York, New York 10016

Oxford is a registered trademark of Oxford University Press

All rights reserved. No part of this publication may be reproduced,
stored in a retrieval system, or transmitted, in any form or by any means,
electronic, mechanical, photocopying, recording, or otherwise,
without prior permission of Oxford University Press.

Library of Congress Cataloging-in-Publication Data
Kyburg, Henry Ely, 1928–
Science and reason / Henry E. Kyburg, Jr.
p. cm. Includes bibliographical references.
ISBN 0-19-506253-1
1. Science—Philosophy. I. Title.
Q175.K983 1990
501—dc20   89-77467

1 2 3 4 5 6 7 8 9
Printed in the United States of America
on acid-free paper

For Sarah Kyburg,
partner in many things for many years.

Preface

Who is the reader of this book?

There are two classes of people for whom it is intended. It covers a relatively broad spectrum of topics in the philosophy of science, and so might be a useful textbook for certain groups of philosophy students. At the same time, many of the views possess some degree of novelty, and professional philosophers concerned with the philosophy of science may find material of interest to them in the volume.

It is not hard to make the case for the philosopher of science. Various novelties will be mentioned in the outline that follows.

As a textbook, the volume is not intended only, or even primarily, for philosophy majors. The student may well come from computer science or cognitive science. He or she is probably not a beginner in philosophy, unless already pretty sophisticated in some science. First-order logic is not absolutely required, but the notation is used rather freely.

On the other hand, no specific presuppositions, regarding philosophy or anything else, are made with respect to the preparation of the reader. It is supposed, first, that the student is what we might call "culturally literate" about science; and second, that the student either has preparation in first-order logic or will be guided through enough of the notation to get along.

It would certainly help a lot if the reader had had at least one philosophy course. It would certainly help a lot if the reader had had at least one serious course in a science.

The original audience for the material in this volume comprised students of cognitive science, nearly all of whom have had some first-order logic, and nearly all of whom have had something of a scientific background.

It should be emphasized that this is in no way a survey of current thought in the philosophy of science. If it is to be used as a textbook for students in cognitive science or computer science, perhaps not much supplementation is needed. For philosophy students, however, a broader background in traditions in the philosophy of science should be provided to put the arguments and views of the present work in a historical perspective. One way to provide this would be by incorporating into the course either a book of readings in the philosophy of science, or a couple of basic background books such as Hempel's *Philosophy of Physical Science*[1] and Kuhn's *Structure of Scientific Revolutions*[2]. There are, of course, many other possibilities, but the basic idea is that from the point of view of the author, the present volume contains the truth about a number of interesting issues in the philosophy of science, and is not intended as a historical review of these issues.

A brief exposition of the contents of the book may be of use to all of the kinds of potential readers mentioned.

Chapter 1 presents a rather informal, free-wheeling characterization of what the philosophy of science is all about, and its relation to science in general and the humanities in general.

Chapter 2 offers a relatively conventional (indeed, conventionalistic) view of the philosophy of logic and mathematics.

Chapters 3 and 4 reflect the author's views on probability and induction. Relatively little attention is given to alternatives, since the author, rightly or wrongly, feels that there *are* no viable alternatives, and the main concern of this volume is not historical balance, but the best information we have.

Chapters 5 and 6 represent the first and most basic applications of the author's approach to probability and induction: observation and measurement. These chapters also provide an extended example of the operation of the epistemic system envisaged.

Chapters 7, 8, and 9 deal with the problem of how we get to hold the general theoretical scientific views we hold. The approach is conventionalistic and pragmatic; it is argued that this is feasible within the framework offered, though it depends on parameters ("levels of accep-

---

[1] Carl G. Hempel, *Philosophy of Natural Science*, Prentice-Hall, Englewood Cliffs, N.J., 1966.

[2] Thomas S. Kuhn, *The Structure of Scientific Revolutions*, University of Chicago Press, Chicago, 1962.

tance") that can only be elucidated later. It will be argued that the choice of such parameters, while it may be "contextually dependent," can be determined explicitly.

Chapter 10 concerns idealization—a matter that has been somewhat neglected in recent work in the philosophy of science. It is argued that idealization is a central and important matter in science. For example, the additivity of extensional quantities is best construed in this way. Idealization is related to another much neglected aspect of scientific inquiry: approximation. It is clearly not the same thing, though both are important.

Chapters 11, 12, and 13 concern causality and dispositions. The general conclusion is that causality is *not* central to scientific inquiry (though clearly very important to engineering) and that modal operators in the object language are not needed in the representation of scientific fact.

Chapter 14 concerns a somewhat marginal matter (from the point of view of the philosophy of science): decision theory. A brief treatment of decision theory is needed, however, since it is invoked in the discussion of the selection of the level of acceptance that is to count as "good enough for practical purposes."

Chapter 15 develops the relation between decision and practical certainty and uses this to resolve the question, hanging open since the early chapters, of how to choose levels of acceptance.

Chapter 16 concerns speculation. It is argued that speculation falls outside of the rational framework that has been developed in the earlier chapters; that it is essential to scientific progress; and that in the realm of speculation "anything goes."

Finally, Chapter 17 explores the limits of science, and the question of whether there are presuppositions or assumptions required by science. It is argued that there are none.

There are more detailed treatments of a number of these issues in articles or in books, though there are also a number of treatments that are new. It nevertheless seems well worthwhile to bring all of these arguments together in one place, since it is the way they function together to give a unified and coherent picture of scientific knowledge that is their real test.

Many people have contributed to the ideas developed here. Discussions with Isaac Levi and with Teddy Seidenfeld, though neither would wholly endorse the conclusions presented here, have been influential in many respects. Students of philosophy, who have sat through courses in which various of these ideas have been tried out, and especially students of cognitive science, computer science, and other disciplines, who have offered arguments and ideas from a perspective different from that developed by and among philosophers, have helped enormously.

It is true that philosophers, over the years, have focussed on a number of interesting and important issues in the philosophy of science. But as in many another discipline, it is the questions and doubts that come from outside that are the most telling. Thus I thank my students from other disciplines, as well as from philosophy; and thus I thank my colleagues from computer science, history, psychology, physics, and mathematics.

It is only at a university in which interdisciplinary activity is as easy as at the University of Rochester that I could have been so influenced by so many people from so many disciplines. I am grateful to the traditions of that university for making my interdisciplinary life easy.

The Center for Signals Warfare of the U.S. Army has provided me, over the past six years, with support for research that, indirectly, has led to a number of the conclusions embodied in this work.

Contents

1. Philosophy and Science, 3
2. Logic and Mathematics, 17
3. Probability, 36
4. Induction, 58
5. Observation and Error, 74
6. Measurement, 94
7. Choosing Among Conventions, 111
8. Laws and Theories, 135
9. Relativity and Revolution, 152
10. Idealization, 168
11. Causality, 182
12. Statistical Causality, 196
13. Dispositions and Modalities, 211
14. Decision Theory, 224
15. Levels of Corpora, 241
16. Speculation, 255
17. The Limits of Science, 265
    Name Index, 277
    Subject Index, 279

SCIENCE & REASON

1
Philosophy and Science

Background

Science—the scientific method, the libraries of scientific knowledge, the sophisticated theories that guide us to the inside of the atom and to the outer reaches of the universe—is the glory of Western culture. Science has given us such power over the world we live in that we are capable of destroying the very culture that gave it birth, but also such power that we are capable of changing the world for the better in a thousand (incompatible) ways. The power of science has opened up new worlds of choices for individuals, for groups, for nations, and for mankind. But as the ancient myths have told us, power bears the potential for evil as well as good; to use our scientific knowledge well, we must understand its basis and its content. We must achieve knowledge of the scope and nature of our scientific knowledge.

Philosophy, that most general of disciplines, can give us the perspective we seek. I have in mind Western philosophy—the philosophical tradition that began in Greece and is still the dominant tradition in Western culture. It is this tradition that spawned science, and thus it should be appropriate to our purposes. But there are many strands of the Western philosophical tradition, many significant philosophical questions to which well-known philosophers have given contrary answers. I intend here no survey of general philosophical perspectives on science, even limited to the Western tradition. Rather, I will select from among those philosophical ideas that seem to have contributed most to our understanding of science, and I will add some novel ingredients. The result, I hope, will be a view of science and of scientific inquiry that exhibits features of coherence and unity that will render it attractive to philosophers as well as to scientists and laymen.

The philosophical tradition of which I speak starts with the pre-Socratic Greek philosophers (some of whom, with a certain amount of license, are sometimes construed as anticipating modern scientific theories; for example, Parmenides is said to have anticipated unified field theory, and Democritus is taken to have anticipated atomic theory). Our scientific tradition is of equal antiquity. Except by the imposition of anachronistic standards, it is hard to fault the work of Euclid in mathematics, Aristotle in biology, or Archimedes in mechanics.

Descartes (of Cartesian coordinate fame) is generally credited with marking the beginning of "modern" philosophy. Descartes focused on the problem of epistemology: What do I know, and how do I know it? The British empiricists, Locke, Berkeley, and Hume, took the second part of the question most seriously and answered that the basis of knowledge is experience. Descendants of their views, among them logical positivism and logical empiricism, have been particularly concerned with scientific knowledge and have tended for many years to dominate Anglo-American philosophy.

Modern science is often thought to have begun in the sixteenth or seventeenth century, more or less coincidentally with modern philosophy. Copernicus and Galileo are given the credit for the great break from the heavy hand of tradition. I shall argue that the contrast between science and philosophy is overdrawn, and that the contemporary academic and administrative distinction between philosophy and science injures both disciplines.

Whether or not it is appropriate to view science as a relatively recent invention, it is certainly true that most of what we now take to be scientific knowledge is a development of the last three or four centuries. It is an accelerating development, and one that has led to such profound changes in the way we live (and in the lengths of our lives) that it boggles the reflective mind. It is in part this mind-boggling quality that leads many people to the philosophy of science: how are we to understand this institution that in a flicker of time (historically speaking), a nanoflicker of time (biologically speaking), has changed the face of the world both culturally and physically?

The idea that might makes right affects our attitude toward science. (Not, I shall argue, completely irrationally!) Science has shown its power, so it must be right. That is, it must be *reasonable* or *rational*. Mathematics is the very paradigm of a logical and rational discipline; much of science embodies mathematics, and all of science emulates it to some degree. The scientific man is the rational man. How do we know? Why, because of his success in predicting the future course of events, in controlling and modifying the world. Because of the *power* of science. But do we call scientific activity rational because of its success, or is it successful because it is rational? Is our assessment of science as rational itself rational and justified? Is scientific "rationality" any different from any other sort of rationality?

The relation between reason and science is more complicated than these questions suggest. Nevertheless, a first rough answer to the last question is that logic (construed broadly enough to include both deductive and inductive inference) is universal; it is the same in science, in common sense, and in philosophy. To show that logic, operating on experience, suffices to justify our corpus of scientific knowledge is another matter. The very claim is controversial. But it is what I seek to show. To the extent that I am successful, there can be a court of reason before which controversial scientific claims — and there are many — can argue their cases.

The Influence of Science on Philosophy

Modern science and modern philosophy have grown up together. Obviously, they have influenced each other. Kant, for example, was motivated by the question of how the human mind could come to know the general laws (the Newtonian laws) that characterize the physical world, in the face of Hume's skeptical arguments to the effect that the past and the future are not connected by *reasons*.[1] Most modern philosophers have felt it necessary to come to terms with the fact of scientific knowledge about the world, though various philosophers have construed that knowledge in various ways, and some, like Bergson,[2] have regarded our physical knowledge as mechanical, dead, and second-rate.

The most optimistic and thoroughgoing responses to the growth of scientific knowledge were to be found in the Vienna Circle and the Berlin Circle in the 1920s and 1930s.[3] Both groups included philoso-

---

[1] David Hume, *An Enquiry Concerning Human Understanding: Selections from a Treatise . . .* Open Court, La Salle, Ind., 1949.
[2] Henri Bergson, *Creative Evolution*, Random House, New York, 1944.
[3] See A. J. Ayer (ed.), *Logical Positivism*, The Free Press, New York, 1959, particularly Ayer's excellent introductory essay.

phers and scientists, and both were deeply influenced by the developments in logic initiated by Frege and by Whitehead and Russell, and by the profound changes in physics that took place in the early part of the century. The Newtonian view of space and time and the Newtonian laws of mechanics were replaced by Einstein's more general treatment. Quantum mechanics, in which randomness replaced the (never-realized) ideal of perfect prediction, was being developed to account for subatomic phenomena. The response was, loosely speaking, logical positivism, a philosophical position characterized more by a core of attitudes than by a list of accepted principles. The positivists were deeply impressed by the progress of science (some might say overimpressed) and reacted strongly against what they took to be the sterility of traditional philosophy.[4]

A contrast that appeared striking to the early positivists was that between scientific progress — new knowledge was accumulating exponentially — and the lack of social progress. One does not have to have settled on any particular criterion of social progress in order to be struck by this contrast. Mere lack of change sufficed to make the point. But particular acknowledged evils — nationalism, prejudice, intolerance, war — were attributed to the "unscientific" use of language, that is, to the use of meaningless words ("glory," "the fatherland," "degenerate"), as though they carried objective meaning in the same sense as "rational number," "six meters," and "velocity."

Metaphysics, for many centuries a respected part of philosophy concerned with deep and general properties of reality, came in for special criticism; indeed, the very word "metaphysical" became a term of opprobrium, roughly synonymous with "meaningless." There was, correctly or incorrectly, a perceived association between the evils of nationalism and prejudice and war, and the metaphysical positions of Hegel, Nietzsche, and other traditional philosophers, as well as the less traditional views of Wagner.

Schlick, Feigl, Carnap, and Gödel in Vienna, and Hempel, Reichenbach, and von Mises in Berlin, imagined the dawn of a new philosophical epoch, in which the business of philosophy consisted in formalizing scientific theories in the new formalism of Whitehead and Russell, analyzing them, and explaining in scientific terms, or explain-

---

[4]In some cases, it was perniciousness rather than sterility that the positivists reacted against. Some brands of traditional philosophy were being used to support nationalism in one of its least attractive manifestations.

ing away, or castigating as "metaphysical" most other traditional philosophical concerns.

The new epoch never dawned. The positivist program ran into a number of difficulties, some of which we shall explore in later chapters. Nevertheless, its influence on subsequent philosophy has been profound, particularly in America and England (where many of the original positivists ended up) and in Scandinavia. A. J. Ayer, who participated in a number of meetings of the original positivistic groups, attributes much of the thrust and many of the specific concerns of the general analytic movement in American and British philosophy to the influence of the positivists.[5] Logical empiricism, primarily an American phenomenon, while never a formal "movement," shared many of the convictions and much of the methodology of the positivists.

It is often said that positivism and logical empiricism are dead. Such pronouncements deserve a certain amount of skepticism, since it is unclear just what it *is* that is dead. But in philosophy it is true that there are few avowed positivists carrying on, and few empiricists regard the original program of the Vienna Circle as feasible. Nevertheless, the methodology of formalization and analysis seems to be alive and well, even among those who periodically write the obituaries for positivism and empiricism. What is gone is the conviction, perhaps due to the early Wittgenstein, that *only* scientific and formal (logical, mathematical) statements are meaningful, the former embodying "content" and the latter being empty but useful tautologies. The remarkable success of physics in the early part of this century misled the positivists into thinking that only science was worth taking seriously.

Thus, particularly in this century, science has had a profound effect on philosophy. The influence has not been entirely benign. One may attribute the excesses of early positivism to a youthful enthusiasm for the wonders and accomplishments of physical science, and particularly for its remarkable rate of progress. But the uncritical enthusiasm has faded, as uncritical enthusiasms do, and contemporary philosophy, while recognizing an obligation to come to terms with scientific knowledge and scientific inquiry, tends to be more cautious. The vestiges of the oversimplified positivistic account of knowledge are to be found, curiously, among scientists rather than among philosophers.

In philosophy, as elsewhere, the pendulum tends to swing past the golden mean. The very idea of progress in science that so impressed

---

[5]Ayer, op. cit.

the positivists is now questioned by those whose concerns are largely historical. Thomas Kuhn,[6] who is influential in contemporary discussions of scientific change, discusses scientific change essentially as a social and historical phenomenon. If there *is* progress, if scientific changes *are* rational and justified, the reasons underlying those changes are hard to find and assess. Without denying that there may *be* rational reasons for change, many contemporary philosophers of science focus primarily on the psychological and social sources of scientific changes.

I do not propose to deny that these psychological and social factors are interesting and worthy of study. But my focus here will be on the *reasons* for scientific change and the justification of scientific knowledge as it exists at any moment. This is not to say that, in ordinary contexts, scientific knowledge stands in need of explicit justification. It is only to point out that there are occasions on which the question of the acceptability or justifiability of some item of putative scientific knowledge becomes an issue. We should be able to address that issue in rational terms.

In this respect, the point of view adopted here is close to that of the old-time positivists. Much of what follows consists of establishing a framework and a way of looking at scientific inquiry that avoids the problems that undermined the cheerful optimism of the positivists. And our efforts are not devoted only to avoiding old problems, but also to providing a framework for the solution of problems that have not even been formulated yet — problems that we can dimly see emerging from the disciplines of cognitive science and artificial intelligence.

The Influence of Philosophy on Science

It would be a large and unrewarding task to try to trace in any detail the influence of philosophy on science. Einstein,[7] for example, waxes philosophical both in his support of relativity theory and in his attacks on the indeterminacy of quantum mechanics. No doubt philosophical considerations played a role in leading him to the theory of relativity, but it is not easy to determine whether these considerations played

---

[6]Thomas Kuhn's most influential book, *The Structure of Scientific Revolutions*, is as familiar to many natural and social scientists as it is to philosophers.

[7]Albert Einstein, *Out of My Later Years*, Philosophical Library, New York, 1950.

merely an accidental role (like the apocryphal role of the apple in inspiring Newton), or a more fundamental (psychological) role in the discovery of the theory, or a yet more fundamental role in the justification of the theory. Similarly, the Copenhagen interpretation of quantum mechanics draws on distinctively "philosophical" arguments, but it is not clear to what extent these arguments are essential to the theory, or to what extent they underlie either the theory's discovery or its justification.

There are situations, however, in which specific philosophical viewpoints influence the methodology and therefore the results of scientific inquiry. An example of this is the prevalence of operationalism in psychology over a fairly extended period. Operationalism, a philosophical approach to science developed by the physicist P. W. Bridgman[8] in 1927, identifies the meaning of a scientific concept with the operations by which we measure it or apply it; the concept, to be meaningful, must be given an "operational definition." (Note the similarities—not coincidental—to the positivistic contrast between "scientific/meaningful" and "metaphysical/meaningless.") Thus "temperature" is given a meaning by the operations we use to measure temperature. In psychology, correspondingly, a concept such as intelligence is given meaning by the operations we perform (the administration of IQ tests, for example) to measure it. There is sound general advice here, but it is also easy to see how the general philosophical position can have pernicious effects in science.

There have no doubt been terms employed in psychology that are so far removed from measurement or experiment that their excision is a good thing. But many terms (the atomic theory may contain examples from physics) may not earn their keep experimentally until some time after they have come to play a significant theoretical role. To demand definitive empirical procedures for the application of every term may inhibit the conceptual proliferation from which novel and deep theories emerge. On the other hand, not every term for which we can give an operational definition is potentially useful. A psychological instrument—a test—may yield consistent numbers that have no important psychological significance. The operational imperative risks rendering the problem of test validity empty: intelligence is just that quality that intelligence tests measure, by definition. And then we must face the question: what is this measured quality good *for*?

[8]P. W. Bridgman, *The Logic of Modern Physics*, Macmillan, New York, 1927.

A more recent example of the influence of philosophical positions on the pursuit of science, again mainly in the social sciences, is Thomas Kuhn's notion of a *paradigm* in science.[9] According to Kuhn, the history of science consists of periods when a particular paradigm — a way of doing experiments, a way of arguing in defense of results, a way of approaching problems — is dominant, separated by revolutions during which one paradigm is replaced by another. The notion of a paradigm is not altogether clear (and Kuhn himself has for some years preferred the more descriptive term "disciplinary matrix"),[10] but this has not prevented historians, economists, and psychologists from coming forth several times a year with proposals of "new paradigms" for their disciplines. Encouraging such grandiose efforts is probably not beneficial to the discipline concerned and is certainly not conducive to philosophical understanding.

A quite different sort of example of the influence of philosophy on science is provided by cognitive science. There have long been cognitive psychologists; since the advent of the digital computer, researchers in artificial intelligence have been seeking to model or improve on human cognitive abilities. In relatively recent years, it has become clear that many of the problems faced by both cognitive psychologists and people working in artificial intelligence either are *philosophical* problems or are so closely related to them that it is helpful to consider the philosophical work that has been done on them. Ironically, some of these very individuals who recognize the importance of philosophy hold naive and oversimplified views on the philosophy of science that make it difficult for them to assess the value of what they find in philosophy. It is not so much ignorance of specific philosophical results that interferes with progress as a failure to understand philosophical method.

Nevertheless, it is becoming clearer and clearer that careful analytic and formal philosophical work can bear directly on the problems faced by cognitive scientists. What is less clear is whether these problems are really scientific, or whether they are strictly philosophical. In the latter case, perhaps we should speak of a new style of cognitive philosophy rather than of cognitive science. Few workers in the field doubt the usefulness of the contributions made by those whose job descriptions

---

[9] Kuhn, op cit.
[10] Thomas Kuhn, "Second Thoughts on Paradigms," in F. Suppe (ed.), *The Structure of Scientific Theories*, University of Illinois Press, Urbana, 1974, pp. 459–482.

include the title "philosopher." But many psychologists, computer scientists, and linguists evince a certain discomfort when talking about the contributions of philosophers—as though, while they are happy to acknowledge the contributions, they would prefer them to have been made by people whose professions are more closely tied to experimental and observational research.

In due course we shall have a lot more to say about experiment and observation, as well as more to say about the formal aspects of scientific theories. For the moment, however, these examples should suffice to show that philosophy, and especially the philosophy of science, is not without its effects—sometimes pernicious, sometimes not—on the practice of science.

The Relation Between Science and Philosophy

What *is* the difference between science and philosophy? We have been acting as though it were perfectly clear, and indeed, no student is likely to be confused about whether he is in the physics department or the philosophy department. Philosophy is one of the humanities, and science is—well, science. In terms of the contrast celebrated by C. P. Snow,[11] we have here two entirely different cultures, the inhabitants of which can barely communicate at all with each other. The humanities focus on human products, on history, on the play of ideas; the sciences focus on the world, on facts, on the accumulation of new, down-to-earth knowledge.

Consider the contrast between medicine and history. History is concerned with facts—we need to know what happened—but what we seek from history is understanding. Our knowledge of the facts may increase, but even if it did not, history would not die. (The acceptance of falsehoods, on the other hand, is potentially devastating!) We can always deepen our understanding, look at the facts from new angles, focus our attention on new historical principles. Medicine is concerned with people and their diseases, but what we seek from medical research is new knowledge of diseases, their mechanisms, and their cures. Without new facts, the *practice* of medicine would continue, but medical science would have ground to a halt.

But consider mathematics. Mathematics is regarded as one of the sciences. But it deals entirely with products of the human imagination

---

[11]In his popular book *The Two Cultures*.

(*pace* Kronecker, who felt that God had created the natural numbers but that all else was the product of man[12]). Mathematics depends on the discovery of new facts, but the facts concern the mathematical structures that people invent. When the old structures become too well understood, when all the interesting theorems about them have been proved, mathematicians invent new structures in order to have something to prove theorems about. So there *are* mathematical facts, and mathematical research depends on the discovery of new facts just as much as does medical research, but the facts of mathematics are facts about human creations, and the point of the facts is to give us a deeper understanding of mathematical structures. What we seek in mathematics, as in history, is understanding of a human product.

The example of mathematics suggests that the honored contrast between the sciences and the humanities is not a constructive one. Philosophy exhibits, within itself, the same polarity of facts and understanding. Since it is often taught as history—we seek to understand the great ideas of the past—it tends often to be "humanistic" in focus. But philosophical research, like mathematical research, seeks to uncover new facts—new theorems—concerning philosophical structures. This is most apparent in logic, traditionally a domain of philosophy, which is so much like mathematics that in some universities it has been relegated to the mathematics department, becoming thereby one of the sciences as opposed to being one of the humanities!

We should therefore not simply accept the lesson of the university catalog that physics is one of the sciences and philosophy is one of the humanities, and that's that. We should look a bit more closely at some of the alleged contrasts between science and philosophy. A number of such contrasts have been drawn, and even more reside, inarticulately, in the perspectives that both those trained in science and those trained in the humanities bring with them as unconscious baggage when they look at the works of the "other side." One of my objects is to exorcise the biases of both "humanists" and "scientists"—many of which they share—in the hope that we can acquire both new facts about and a new understanding of the enterprise of science.

Let us be more precise about the relationship between science and philosophy. The most obvious difference to the layman is that while philosophy still chews on the same perennial problems that bothered

---

[12]As quoted by many, including R. L. Wilder, *The Foundations of Mathematics*, John Wiley, New York, 1952, p. 192.

Plato and Aristotle, science progresses by leaps and bounds. But this is a misleading way to put the matter. Aristotle worked on biology and physics, as well as on ethics and logic. Indeed, all knowledge was the domain of philosophy until well into the Middle Ages. What we see, rather than the development of new scientific disciplines from scratch, is the isolation from general philosophy of a set of problems and a set of techniques for approaching them, and the corresponding development of a group of people specializing in those problems and techniques.

What motivates this specialization is that applying these techniques to these problems yields *results*. In many cases, a large part of the inspiration and motivation comes from technology, long regarded as being immeasurably beneath philosophy. Thus another way of looking at progress in science and lack of it in philosophy is to see philosophy as concerned with the (remaining) *hard* problems. Once a relatively uncontroversial approach to a certain set of problems has been worked out, some people specialize in dealing with those problems and the true philosophers turn to other domains.[13]

Two things are worth noting in particular. First, the same phenomenon occurs within the classical sciences: we see chemistry dividing into organic and inorganic, and biochemistry splitting off from organic chemistry. Second, the opposite phenomenon also occurs. The world is all of a piece, and problems do not arise neatly categorized into the conventional academic disciplines (physics, chemistry, psychology, etc.). Psychology, neurology, physiology, and biochemistry, for example, are all involved in understanding vision. But this means that what practical exigencies of research have put asunder must, for the study of vision, be knit together. Sometimes (as in the case of cognition—i.e., for another set of practical exigencies) this may even require the consideration of problems that fall outside any of the relevant specializations: that is, hard problems, problems still within the domain of philosophy.

The foregoing also suggests an explanation of why people tend to think of science as achieving *results*, while philosophy gets nowhere and only chews on its own tail. Most of the "results" of science are small: we extend our knowledge of Young's modulus for a certain steel alloy a few decimal places, we refine a reaction-time experiment to

---

[13]Science also evolves from technology, but I think it does not become what most people would call *science*, as opposed to lore or engineering, without some admixture of the abstract and general concerns of philosophy.

achieve slightly more uniform results. But small results are to be found in philosophy too: we discover the hidden premises that are required to make an argument go through, we reveal the ambiguity that gave an invalid argument the appearance of validity. Philosophers make mistakes, they get caught out, and the same mistakes are not repeated.

But some results in science are large, and there seem to be fewer large results in philosophy. Partly, this is an illusion: a large result in philosophy often leads to the establishment of a new scientific discipline, and the result is therefore associated with the science rather than with philosophy. And partly, the relative paucity of large results in philosophy is just a fact. Real breakthroughs in philosophy are rare. But we already noted that philosophy was the discipline that dealt with *hard* problems, didn't we?

So what *is* the difference between science and philosophy—or, more generally, between the sciences and the humanities? A number of contrasts have been drawn. It is sometimes said that science deals with *facts*, while philosophy deals with *values*. But there is ambiguity in "deal with," as well as in "facts" and "values." In one sense, economics is the discipline that deals most directly with values. It deals with values in a descriptive way, though, and it could be claimed that philosophy deals with values normatively. Again the contrast is easy to muddy. Philosophy also seeks descriptive and logical facts about values, as well as about a wide variety of other things. And a philosophy that is not factually informed is not likely to be very relevant to the world as we know it to be.

It is true that, in a number of respects, philosophy is a normative discipline; so is mathematics: logical and mathematical standards of validity are of a piece. And psychology, political science, and economics have their normative aspects. In engineering (shall we regard that as part of science?) normative considerations abound: the best way to accomplish A is by means of X. And "pure" science has its own imperatives: accuracy, honesty, and, of course, purity. But it also has less global imperatives: the first thing to try is to add weak hydrochloric acid; you don't really understand the ecological role of Y unless you know who eats Y and who is eaten by Y.

There may be interesting distinctions to be drawn among the various ways in which the various disciplines deal with facts and values, and the ways in which various disciplines function normatively. No doubt there are. But while these distinctions may be informative and enlightening philosophically (or sociologically), they are not sharp and obvious enough to serve the purpose of dividing the intellectual activities on a campus into two kinds.

How about observation and experiment? Leaving aside mathematics, observation and experiment seem to characterize the sciences and to play no role in philosophy or the other humanities. Of course, even in the most experimental sciences, most of the work consists of designing experiments and in analyzing their results. The actual conduct of experiments is often left to laboratory assistants or to graduate students. And in the more theoretical sciences, and in the more theoretical parts of any traditional scientific discipline, scientific work consists of proving theorems, exploring the literature, devising and analyzing arguments, and imagining claims or axioms or explanations that might, in some remote fashion, be ultimately put to the test. Much of this involves activities and methodology indistinguishable from those employed in philosophy.

"Ultimately, however, science rests on sense experience." So speaks the relatively sophisticated physicist. Perhaps. This is a view that has been explored in philosophy, notably by the British empiricists Locke, Berkeley, and Hume. Carrying the argument to its extreme, Berkeley[14] argued that the world consists only of *ideas* (mental entities) and that the very notion of a material substance is incoherent. This is a conclusion that would probably be unwelcome to the physicist.

But we understand what he means, and many of us sympathize. As a loyal descendent of the positivists and as a (sort of) practicing logical empiricist, I, too, believe that knowledge is ultimately based on experience, though it must be confessed that what we know has an effect on what we experience, and that what we are genetically determines what we *can* experience. Nonetheless, in some sense, what we can rationally claim to know surely depends on what happens to us.

But this is no more true of what are conventionally called sciences than it is of literature or history. If our friend the physicist wishes to infer that all rational knowledge is part of science, he may do so. But we thereby lose some distinctions that are of more than administrative significance.

---

[14]Bishop George Berkeley, a famous British empiricist, denied the independent existence of matter on roughly the same grounds that incline many to deny the existence of the tooth fairy: we have lots of evidence from our senses that teeth are replaced by dimes, but none that this is done by a tooth fairy; we have lots of evidence from our senses that sensations succeed each other in ways that are often orderly and predictable, but no sensible evidence that this succession has anything to do with matter.

A Better Classification

One of the difficulties in assessing the relation between science and philosophy stems from construing philosophy as one of the humanities, and therefore as a kind of opposite pole from experimentally based science. One way to alleviate the polar effect is to introduce a three-way distinction to replace the polar distinction. In fact, I think that the following distinction does more, and I commend it to the attention of university administrators.

Let us distinguish academically among *formal* disciplines, *empirical* disciplines, and *interpretative* disciplines. Mathematics is a formal discipline, biology and psychology are empirical disciplines, and literature is an interpretative discipline. It should be immediately clear that every actual discipline embodies aspects of all three types: much of mathematics is ultimately tied to facts about the world; biology is concerned with formal structures on occasion, and psychology involves interpretation; literary criticism deals with both the formal structure of the poem and the facts about the society that produced it.

In this framework, philosophy, like mathematics, is essentially a formal discipline, and the sciences we are primarily concerned with, like biology and physics, are largely empirical. Our formal concerns have to do with the structure of scientific knowledge and of scientific theories, the formal relation between experimental or observational results and the acceptance of scientific theories, the formal development of theories of measurement and error, and the like.

We could also attend to an interpretative aspect of both science and philosophy. Scientific theories arise in a certain milieu and against a certain background of philosophical thought. Understanding a unique event in the history of science is a very different matter from analyzing the formal relations that obtain between a newly emerged theory and the experimental data by which it is alleged to be supported.

Both approaches are of interest, and each may well be, in some sense, essential to the success of the other. But here we will follow the formal route, in keeping with the view that philosophy is primarily a formal discipline. I take the interpretative approach to belong more to history than to philosophy.

Formality and insight are not exclusive, however. Indeed, we hope that the relatively formal exercises that follow provide insight both into the role of reason in science and into the role of science in the life of reason.

2

Logic and Mathematics

Mathematics is traditionally regarded as one of the sciences, despite the fact that one is hard put to find an empirical basis for it. J. S. Mill[1] suggested that the law that $5+7=12$ is an empirical generalization of the fact that when you combine five things and seven things and count the result, you (almost always) get twelve. As A. J. Ayer has pointed out,[2] if I count five pairs of objects and get nine, I do not regard that as casting doubt on "$2 \times 5 = 10$," but as evidence that I have counted wrongly. Nagel notes[3] that "$5+7=12$" doesn't work for combining gallons of water and gallons of alcohol. In fact, few people have been convinced that arithmetic laws are on a par with those of mechanics or chemistry. With respect to geometry, more people have been convinced that we are concerned with a very general physical theory. It is said, for example, that Einstein has taught us that "physical space" is not Euclidean, after all, but "curved." And some people, for example Hilary Putnam,[4] have gone so far as to say that even logic is not immune to modification in the light of physical discoveries; he has proposed that a nonclassical logic be employed in quantum mechanics.

These are clearly issues that we should consider in the philosophy of science. But there is another reason for the present chapter, and that is that logic (in fact, classical logic) will provide the framework and ground rules for our subsequent analyses. Scientific theories will be

---

[1] J. S. Mill, *A System of Logic*, Longmans, Green, London, 1949, Book II, Chapter VI, Section 2. (Original publication, 1879.)

[2] A. J. Ayer, *Language, Truth, and Logic*, Dover, New York, n.d. (Original publication, 1934.)

[3] E. Nagel, *Logic Without Metaphysics*, The Free Press, Glencoe, Ill., 1956, p. 84.

[4] Hilary Putnam, "Is Logic Empirical?" in R. S. Cohen (ed.), *Proceedings of the Boston Colloquium in the Philosophy of Science*, Humanities Press, New York, 1969.

expressed in a first-order extensional language, and we will talk *about* them in a corresponding metalanguage. Both the character of the object language and the character of the metalanguage should be described here, for these are our tools for subsequent work. It is also important to present some argument for avoiding excursions into intensional and modal logics, and for minimizing the semantic component of our inquiry in favor of the syntactic component.

A third reason for spending time on logic and mathematics is that logic and mathematics are among the most striking and characteristic human creations. If we are concerned with understanding human reason at its best, these formal disciplines are paradigmatic. This is not to say that human reasoning—the mental process—operates formally. There is some psychological reason to think that it does not.[5] A case could be made that knowledge of formal procedures is an aid to rational thought. Below we shall argue that the formal structures of logic and mathematics provide an important framework for communication.

Even those who deny that human reason works by formal rules, and even those who deny that ideally it ought to work this way, will agree that it would be desirable to arrive at the same conclusions, in many cases, that are arrived at by formal procedures. And the ability, in many cases, to settle questions of the validity of an argument definitively, is a valuable ability.

We need not, I think, spend much space and effort on questions in the philosophy of logic or the philosophy of mathematics, though a few such questions will be treated later on. What does concern us here is the relation between mathematical (and logical) theories, on the one hand, and empirical scientific theories, on the other. This distinction may not always be easy to draw, and it may not even be a tenable one in the final analysis, but there are simple cases in which it *seems* quite clear. Abstract algebra is about certain kinds of purely abstract objects, defined to satisfy certain abstract constraints. Mechanics is about forces and pulleys and levers, many of which we can instantiate

---

[5]Ryan D. Tweney and Michael Doherty, "Rationality and the Psychology of Inference," *Synthese* 57, 1983, 139–162, cites empirical evidence to the effect that rational people do not follow the canons of formal logic in inferring. L. J. Cohen, in "Can Human Rationality Be Experimentally Demonstrated?" *Behavioral and Brain Sciences* 4, 1981, 317–331, maintains that since the canons of logic just codify what the best people do, this is almost self-contradictory. Cohen's paper also contains an extensive bibliography on the issue.

in the world around us. But forces and pulleys and levers, as treated in mechanics, are idealized objects. And such down-to-earth things as the numbers on bank statements do in fact instantiate the structures of abstract algebra. In some respects, our discussion of these matters must be limited and anticipatory, because the main uses of mathematics in the sciences are related to measurement, and measurement is a topic we will not get to for several chapters.

Furthermore, we shall argue, in due course, both that scientific theory is rather more a priori than is typically supposed, and that there is a sense in which there are alternatives to classical logic and classical mathematics, and that the choice among these alternatives reflects the exigencies of human communication and human thought (particularly the former!), as well as the constraints imposed by the course of our experience. But these are themes that we must work up to gradually.

First-Order Logic

Otherwise known as the *predicate calculus*, first-order logic is the least controversial part of logic (barring the relatively uninteresting sentential calculus).[6] This is not to say that it is not a source of controversy. Some writers think that the predicate calculus includes principles that it should not include [e.g., Birkhoff and von Neuman[7] reject the distribution principle $(p \& q) \vee (p \& r) \rightleftarrows p \& (q \vee r)$, for the purposes of quantum mechanics], and quite a few writers believe that it is not strong enough to represent a lot of important relations, such as those ordinarily captured by statements mentioning necessity or causality—that is, that it fails to include what it should include. But so far as it goes, most people go along with it. I shall include identity theory (and definite descriptions, as handled, for example, in Kalish and Montague[8]) as part of first-order logic. I assume that most readers have some familiarity with this logic; the function of these few pages is to make explicit the formal framework we are presupposing.

---

[6]Some familiarity with this part of logic is presupposed throughout the book. There are many good textbooks, but one that allows for easy generalization is D. Kalish and R. Montague, *Techniques of Formal Reasoning*, Harcourt, Brace and World, New York, 1964 and later editions.

[7]G. Birkhoff and J. von Neuman, "The Logic of Quantum Mechanics," *Annals of Mathematics* 37, 1936, 823–843.

[8]Kalish and Montague, op. cit.

First-order logic applies to a language consisting of individual variables ($x, y, z, \ldots$), predicates taking zero or more arguments ($S, P, Q, \ldots$), operators taking zero or more arguments ($M, N, \ldots$), logical constants, punctuation, and so on. Fixed sentences can be represented by zero-place predicates, proper names by zero-place operators.

We may formalize first-order logic as applied to such a language by providing a syntactical characterization of its *terms* and *formulas*. A *term* is a variable or an *n*-place operator followed by *n* terms or a variable binding operator followed by an appropriate expression, such as a definite description. A *formula* is a predicate followed by an appropriate number of terms; or a pair of terms flanking "="; or a conjunction, negation, and so on, of a formula; or a quantification of a formula. The logical constants are given by enumeration.

A number of axioms, or axiom schemata, are provided for the logic, again by enumeration. The use of schemata allows us to give an infinite number of axioms in a finitary way; for example, we can say that every formula of the *form* $\ulcorner(x)(S \to T) \to ((x)S \to (x)T)\urcorner$, is an axiom.[9]

Finally, a number of rules of inference are provided—for example, *modus ponens*: from S and $\ulcorner S \to T \urcorner$ to infer T.

There is a system; what about it? First, note that we have already been talking *about* the system. That is, we are using English (supplemented by a few other expressions functioning as *names* of expressions in the object language) as a metalanguage to talk about this formalized language which is to embody first-order logic.

There are other things we can say, also in the metalanguage, also requiring no more than syntax. For example, we can define the notion of *proof* (and therefore of *provability*): a sequence of formulas $F_1, F_2 \ldots F_n$ constitutes a proof of a formula G just in case each formula in the sequence is an axiom or is the result of applying a rule of inference to one or more formulas earlier in the sequence. With a little attention to variables, we can go on to define "derivable from premises" in a similar way, and of course, a formula that is provable from no premises is just a theorem. We may be able to prove that this system is consistent—that is, that it is not possible to derive a contradiction as a theorem.

---

[9]Whereas quotation marks are used to form the names of specific expressions ("a is a cat."), *corners* are used to form ambiguous names of *kinds* of expressions ($\ulcorner Px \to Qx \urcorner$).

We may represent systems in first-order logic that go beyond first-order logic itself. For example, we could introduce predicate constants "*Pt*," "*Ln*," "*Lies-on*," and some axioms and have a system of geometry. We are *construing* these new predicates as geometrical properties and relations, but we are not leaving the realm of syntax. If we want to provide a formal *interpretation* of these predicates, we will have to be concerned with the semantics of our system.

Another way of looking at the formal first-order system is semantically. Let a *model* for this language consist of a universe U of objects; to each variable and proper name, assign some object in U; to each n-place predicate, assign a subset of $U \times U \times \ldots \times U$ (the n-fold Cartesian product of U with itself); to each n-place operator, assign a function from $U \times U \times \ldots \times U$ to U, and so on. We can now, following Tarski's approach,[10] define *truth in a model*: the sentence Pa is true in a model just in case the object assigned to a belongs to the subset of U assigned to P, and so also in the more general cases.

With the notion of truth at our disposal, we can say some really substantive and exciting things. We can show (for a standard first-order predicate calculus) that all the axioms are true in all nonempty models. We can prove that the rules of inference preserve truth in a model, and therefore, that the theorems are true in all models. We can show that proof from premises has the property that the conclusion will be true in any model in which the premises are true. This is *soundness* and corresponds to the intuitive notion of a valid argument: one in which, if the premises are true, the conclusion *must* be true.

Furthermore, in this semantic framework we can establish *completeness*: every formula that is true in all nonempty models is a theorem of the system — that is, provable. And if C is true in every nonempty model in which $P_1, P_2, \ldots P_n$ are true, then there is a proof of C from premises $P_1, P_2, \ldots P_n$.

In short, we can show the system to be sound and complete.

The system that embodies some geometry is not true under every interpretation in every domain, like first-order logic itself. So we provide a semantic interpretation for it in a special domain designed for that purpose, such as the following: "*Pt*" is interpreted as a pair of real numbers; "*Ln*" is interpreted as a set of pairs of real numbers satisfy-

---

[10]Alfred Tarski, *Logic, Semantics, Metamathematics*, Oxford University Press, Oxford, 1956. Includes the original: "The Concept of Truth in Formalized Languages," pp. 152–278.

ing a linear equation; and so on. Under this interpretation, the Euclidean axioms become true.

Alternatively, we can interpret, "Ln" as the path of a light ray in a vacuum, "Pt" as the intersection of two light rays, and so on. And then the Euclidean axioms become (according to general relativity) false.

But what do these terms and formulas represent in the general case? What can we *do* with this system? What is it good for? How is all this related to our normative use of the canons of logic? If we suppose that our formal language—the object language—represents a regimented fragment of ordinary discourse, then validity in the formal system corresponds to validity in that fragment of ordinary discourse.

This is to treat the general formal language as interpreted, just as we treated the geometrical language as interpreted. The predicate "P," for example, may be interpreted as corresponding to the English phrase "is red." The supposition involved is that the world is a model of the formal system. This is not altogether innocuous: it requires, for an interpreted language, that there be some subset of the universe or the world that corresponds *exactly* to the set of things to which the English phrase "is red" applies.

This requirement may seem severe and unnatural. It rules out, for the purposes of logic, the very possibility of vagueness and imprecision. But vagueness and imprecision, as we shall see later, are crucial to the productive use of scientific language. Many writers, the best known of whom is Lotfi Zadeh,[11] take this as grounds for preferring a logic that is designed to deal with vague or fuzzy terms.

On the other hand, for settling questions of validity, it doesn't matter in the least what specific subset of the universe of discourse U is taken to correspond to the English phrase "is red." All that is required is that within the context of a specific argument (proof), the interpretation of "is red" can be assumed not to change. (Otherwise, we really are at sea in knowing how to take an argument!) Since validity holds for *all* models, logic is justified in terms of truth and truth preservation.

But for determining the truth of such sentences as "Pa"— corresponding to "This flower is red"—it may matter what subset of the universe corresponds to "red," because then we may need criteria for membership in the set, and those criteria may be hard to find. And of course, we then have thrust upon us the problem of knowing wheth-

---

[11] Lotfi Zadeh, "Fuzzy Logic and Approximate Reasoning," *Synthese* **30**, 1975, 407–428.

er the criteria are *satisfied* or not, and this is just the same problem over again. It will be argued, in due course, that the locus of this problem is *judgment*; we need to know when we can reliably judge whether or not something is red.

As we will see, this problem cannot be resolved without recourse to considerations of probability and uncertain inference. This takes us far afield from the concerns of the present chapter. For our present purposes — considering classical logic and mathematics — the earlier observation suffices: it doesn't matter where we draw the line between red and nonred in our semantic theory; *wherever* we draw it, classical logic will characterize our valid inferences.

One may be disturbed by a feeling of circularity here. For a regimentation that maps a fragment of ordinary language onto the first-order logic to be correct, it must be such that the semantics makes the world a model. But in order to specify the interpretation that makes the world a model in the intended sense, we must already (e.g., in our metalanguage) have the means for specifying the appropriate objects in the world (the set of objects satisfying the predicate "is red").

There is an alternative approach to truth that may be more appropriate to interpreted formal languages. This is Quine's substitutional approach.[12] According to this approach, what is important about $(x)(Fx \vee -Fx)$ is not that it is true in every nonempty model of our formal language, but that it is true no matter what predicate of *our* language we substitute for F. This is the reflection of an older notion: that the validity of an argument is determined by its "form" alone. That notion, in turn, is tied to considerations of ordinary language.

Quine focuses on translation, taking it to be a principle of charity and clarity that our translation of a foreign tongue be such as to attribute to the foreigner the same logic that characterizes our own grammar. This is sensible, and relevant to the philosophy of natural language. But it does not fit comfortably with Quine's conservatism when it comes to first-order logic. Quine finds standard first-order logic in English only by ignoring much of English discourse altogether (all but the declarative part), by issuing promissory notes for more (the part concerning propositional attitudes), and by bending the remainder to his will (e.g., replacing tenses by dates to yield eternal sentences).

---

[12] W. V. O. Quine, *Philosophy of Logic*, Prentice-Hall, Englewood Cliffs, N.J., 1970.

But this way of putting the matter is too harsh. Our concern is not with discourse in general; rather, it is with the already somewhat regimented discourse to be found in the practice of science. For our special purposes, an even more high-handed approach seems perfectly acceptable. In those specialized parts of English that concern physics, sociology, mathematics, psychology, and other disciplines, there are certain predicates, operators, and so on that carry the main burden of the relevant discourse. Rather than attempt to find and characterize the form of those arguments we consider valid in their natural habitat, we may devise a new and artificial language—first-order logic with an appropriate lexicon—into which we can translate the ordinary and extraordinary scientific English idiom with a minimum of loss. The semantics that goes along with this artificial language is exactly such (by design) as to support Tarski's notion of truth (just described) and the usual semantic notion of validity.

What is the point of doing this—of making this departure from conventional and conversational English? Surely it is not to conduct the business of science in the austere artificial language. It is rather that the existence—even the *potential* existence, since few domains are worth completely axiomatizing—of a first-order formalization guarantees that disputes or confusions about validity can be definitively settled. It isn't that one ever has to go as far as presenting a formal proof of the validity of the argument; it's that there exists, within this framework, a normative standard of validity. (But it may well be that one should, in general, go further than one does. My impression is that in artificial intelligence, people tend to get away with murder just because the people they are talking to don't want to be "picky.")

We have claimed that the object language embodying our scientific knowledge may be taken as extensional. This precludes incorporating such operators as necessity and possibility (or, for that matter, provability) within the object language itself, and also precludes incorporating a causal relation among the relations of the language. One of the motivations for intensional logic, many-valued logic, and so on is exactly that the extensional regimentation is felt to be unnatural and overly restrictive. Semantics can be given for these nonstandard logics, too, but it will be a different semantics, often a semantics that invokes possible worlds and relations among them. It is incumbent on us, in due course, to show that the functions served by the special logics either are inessential or can be served in some other way.

The upshot is this: we will represent (hypothetically and ideally) fragments of science in an extensional, first-order language. The lexicon of this language will contain appropriate terms, predicates, and operations, and we will specify certain formulae as axioms. Thus for mathematics (generally included in our object language, whatever the specific subject matter may be), we may include the relational predicate "∈," denoting the membership relation, and appropriate axioms to govern it. We may also, of course, include eliminable terms, predicates, and operators, such as "3," ">," and "+," together with the corresponding definitional axioms. The distinction between logical axioms, definitional axioms, and axioms characterizing the subject matter of the discipline concerned, may not always be easy to make. It may not always be of importance, either.

Metalanguages

In describing a first-order language, we make use, of course, of a metalanguage—usually ordinary English with a few special notations. But there are things to be gained by formalizing the metalanguage as well. Provability, for example, is an important notion, and provability in an object language is most conveniently characterized metalinguistically. More importantly, I have claimed that probability is analogous to provability in being most usefully construed metalinguistically: in particular as a function from sets of sentences of the object language (representing bodies of knowledge or rational corpora) and sentences of the object language to (not necessarily proper) subintervals of [0,1]. Probability will be enlarged upon in the next chapter.

It will thus be necessary that the metalanguage have facilities for referring to real numbers, or at least rational numbers. In general, we may as well suppose that our metalanguage includes the object language as a proper part, as well as containing names for the expressions of the object language. We shall employ quotation marks to form the names of specific expressions of the object language and Quine's device of quasi-quotation to form generic names of types of expressions in the object language. Thus "$2<5$" denotes a specific (true) arithmetic assertion in the object language, and $\ulcorner 2<a \urcorner$ denotes any expression such as "$2<x$," "$2<y$," and so on, where a has been replaced by a variable of the object language.

The first-order languages we shall be concerned with are extensional: the replacement of a term in a complex expression by a coreferential term turns truths into truths and falsehoods into falsehoods, and the replacement of a sentence by one of like truth value in a complex statement similarly preserves the truth value of the original. Many statements in ordinary scientific English are not extensional. For example, consider statements of provability: "'9>5' is provable" is true, but despite the fact that "9" and "the number of books on my desk" are coreferential, "'The number of books on my desk>5' is provable" is false. Similarly for necessity: "Necessarily Tully=Tully" is true, but "Necessarily Tully=Cicero" is not, despite the fact that (in this context) "Tully" and "Cicero" are coreferential. Similarly for causation: "That the car ran out of gas caused it to stop" might be true, but the substitution of other equally true "the car is green" for "the car ran out of gas" would yield a falsehood. Similarly for counterfactuals. Similarly for attitudinals, such as belief, knowledge, doubt, desire, and so on. Similarly for verbs of intent, such as "seek": to seek a unicorn is not the same as to seek a griffin, though the two terms are coreferential.

Many writers take such examples as these as clear (and even conclusive) evidence that the language of science cannot be extensional. Certainly, it is claimed, the *logic* of science cannot be extensional.

In claiming, therefore, that an extensional first-order logic will suffice, I am making a serious and controversial claim. It is one that I cannot support by a short argument. In fact, a fair part of this book can be construed as an attempt to support the claim by exhibiting an extensional approach that deals successfully with a number of issues in the philosophy of science. But I can give a brief indication of the approach that I think will work.

Provability, as already remarked, is not extensional. But we can define it (we have already done so informally) in a metalanguage. Furthermore, since we can define provability in purely syntactical terms, we can do this in an extensional metalanguage. This is commonplace and traditional in treatments of pure first-order logic; "provable" is not a predicate in the object language, but it is a predicate of the metalanguage that is applicable to certain sentences of the object language. Similarly, I construe "probability of" as a functor in the metalanguage, rather than as a functor in the object language.

But what of necessity, belief, causality, and the others? I shall argue in due course that these, too, in many of their contexts, can be construed as metalinguistic notions and thereby can be rendered extensional. The key notion, in most cases, is that of a rational corpus, which we will construe as a set of sentences of the object language.

Ordinarily and initially, we are to think of it as the set of sentences reasonably accepted (but more about rationality anon) by an individual. We can also think of the set of sentences accepted by any of a group of individuals (e.g., the rational corpus that is the union of the rational corpora of contemporary physicists) or the set of sentences accepted by each of a group of individuals (the intersection of the rational corpora of contemporary physicists) or the rational corpus of an individual, purged of "particular" statements, or even the limiting rational corpus of a scientific discipline or of God. We can consider rational corpora generated from various sorts of bases, subject to various sorts of closure conditions, and so on. For example, the empirical part of realistic bodies of knowledge should admit of finite axiomatization. It is common to suppose that rational corpora should be consistent and deductively closed, though in fact there are reasons for making neither assumption. We can even consider hypothetical rational corpora that belong to no real individual. We have in the notion of a rational corpus a very powerful and flexible tool.

To give a brief indication of how this notion can be used, let us look at some traditional puzzles.

"If the match is struck, then it will light." Construed as a simple conditional in the object language, this sentence will be true if the match is not struck (even if it is soaking wet) or if it is thrown into a fire rather than being struck. Construed epistemically, however, what the sentence comes to is that the addition of the antecedent to our rational corpus will warrant the addition of the consequent. It does so in virtue of the knowledge in our rational corpus that struck matches generally light *and* in virtue of the *lack* of any knowledge about this particular match — its being soaking wet, for example — that would keep it from lighting. If we know, of this particular match, that it is wet, then, construed epistemically, the conditional is false. Adding the antecedent to our corpus will not warrant the addition of the consequent.

In artificial intelligence the problem of dealing with such conditionals is called the "qualification problem." In order to know the results of an action, such as striking a match, there is an unbounded list of qualifications that are being made implicitly (e.g., that the earth's atmosphere still contains oxygen immediately after the act of striking the match).

"If the match *had been* struck, it *would have* lit." This statement becomes: make the minimal deletions from your rational corpus that will allow you consistently to add the sentence "The match is struck." Then that corpus will contain the statement "The match lights." "Minimal" needs explanation but — it is claimed — could be given a syntactical characterization; "consistent," although it is a semantic notion, can for practical purposes often be replaced by the syntactical, proof-theoretical notion of not yielding a contradiction; and the closure conditions that lead from the inclusion of "The match is struck" to the inclusion of "The match lights" will also be construed as straightforwardly extensional.

John believes that Tully = Tully but not that Tully = Cicero: "Tully = Tully" is in John's rational corpus, but "Tully = Cicero" isn't. We don't expect substitutivity of identity inside quotation marks. (Compare: cat = domestic feline, "category" ≠ "domesticfelineegory.") What if John speaks only Russian? That doesn't prevent us from *representing* his rational corpus in English. If we are concerned with his *linguistic* beliefs, then the language matters and new complications arise. But we may still be able to handle things extensionally through the object language/metalanguage device. At any rate, the simplicity of this approach warrants pushing it as far as we can.

"Object language." "Metalanguage." "Extensionality." These are terms of art, more or less clear to those with training in logic or philosophy but perhaps obscure to others. The distinction between object language and metalanguage is one that is not enshrined in natural language, but it is a straightforward one. In the present context, the object language is the language of science: the (idealized) language in which our scientific theories and laws may be expressed. Nothing mysterious here.

So why a *meta*language? Because although there are formal languages in which self-reference is possible, and many of the things that we would use a metalanguage for can be done in the object language itself, matters are much simplified by the introduction of a distinct language for these purposes. What purposes? We may wish to talk about *truth* (though I shall claim that we have little reason to), about *theoremhood*, about the *validity* of arguments, about *probability*, about *implication*, and so on, and all of this is more easily done (and, in some cases, perhaps only done) in a language distinct from and richer than the one whose sentences we are concerned with. Thus we not only want to be able to prove that 2 and 3 are 5 in our object language, we want to be able to say that it is a *theorem* that 2 and 3 are 5, while the fact that twice the number of satellites of the earth and 3 are 5 is *not* a theorem.

We have just seen two ways to do this in our English metalanguage: by forming a name of a sentence by means of quotation, and by forming a nominal clause that can serve as the subject of which " . . . is a theorem" can be predicated. Using quotation is employing a metalanguage. Thus we can say that sentence S implies sentence T in our object language L just in case there is a sequence of sentences in L, each of which is a theorem of L, or S, or can be obtained from earlier sentences in the sequence by the official rules of inference of L, and which contains the sentence T. "S" and "T" here are metalinguistic variables that can be replaced by the names of sentences of the object language.

Just as we have defined implication (we take entailment or implication to be a semantic relation between sentences), so we can define theoremhood in terms of provability. It is a property of sentences in our object language. Probability and various other notions can be handled similarly. One of the advantages of defining these notions in a metalanguage is that it is easier to keep our object language extensional.

Exactly what shall we understand by "extensional" in this context? Two things: (a) two sentences will have the same truth value if one is obtained from the other either by the replacement of one sentential component by another having the same truth value or by the replacement of one term by another coreferential term and (b) two terms will have the same reference if one is obtained from the other either by the replacement of one sentential component by another having the same truth value or by the replacement of one term by another coreferential term.

Illustrative examples: "The first coin toss in New York State in the year 2099 will yield heads and $2+3=5$" has the same truth value (whatever it may be) as "The first coin toss in New York State in the year 2099 will yield heads and $5+7=12$," since the components "$2+3=5$" and "$5+7=12$" have the same truth value. But also, for all we know, "The first coin toss in New York State in the year 2099 will yield heads and the second toss will yield tails" may also have the same truth value as the first sentence quoted. Two sentences may obviously have the same truth value without our knowing it!

"The sum of the even prime and 5 is 7" has the same truth value as "The sum of 2 and 5 is 7," since "2" and "the even prime" are coreferential.

Now note that implication, theoremhood, provability, and the like would not be extensional notions in the object language if we tried to represent them there. (That there are five books on my desk implies that the number of books on my desk is odd, but $2+2=4$, which has the same truth value, does not imply that the number of books on my desk is odd.) The same is true of a large number of epistemically important ideas: evidence (that S has the same truth value as T doesn't mean that S supports whatever T supports), belief (e.g., beliefs about the morning star and the evening star), knowledge (since it is usually supposed to require belief), probability (not all true statements have the same probability), and so on.

By eliminating such notions from our object language, we simplify it greatly. By representing them in the metalanguage, we can still make use of them, but now in an extensional context, where, for example, *implication* is a perfectly ordinary relation between terms: "the number of books on my desk is five" bears the implication relation to "the number of books on my desk is odd." Where we construe implication metalinguistically as a relation between *sentences*, there is no violation of extensionality. It is less conventional to treat probability, belief, and other epistemic notions as metalinguistic, but we shall follow that course as well.

To sum up: the metalanguages corresponding to the various object languages with which we shall be concerned will be assumed to be extensional, to include as a proper part the object languages they concern, and to have the expressive power to formulate statements of epistemic probability. This requires that they include some of the machinery of set theory and some mathematics.

Arithmetic

Although it may be construed in other ways, I shall here construe mathematics as the theory of membership, on the grounds that we can construct all the mathematics we need from set theory. But we could also, if our interests were more parochial, consider special mathematical theories independently of set theory. Thus if in a certain part of science our interest extends only to the arithmetic of natural numbers, we could axiomatize this directly, in the fashion of Peano. The principle of mathematical induction must be presented as a schema corresponding to an infinite list of axioms—there is no finite first-order axiomatization of arithmetic—but that presents no problem in principle.

It is also the case that not all arithmetic truths are captured by any set of axioms; arithmetic is essentially incomplete. This is also true of set theory, of course. But for the purpose of considering a possible axiomatization of a fragment of empirical science, we need not concern ourselves with this problem either. All we require is that we have axioms sufficient to warrant those arithmetic transformations and arguments that we actually need in that bit of science. The incompletability of the theory of natural numbers (and anything stronger than that) is of great importance in the philosophy of logic and the philosophy of mathematics, but it has only a marginal bearing on the philosophy of physics.

We should also note that this direct procedure fails to give us a "definition" of natural number. "0" is just a primitive zero-place operator; "N" is merely another primitive predicate in our language. In this respect, the axiomatization in terms of set theory may seem more enlightening, for we can define a certain set of objects that correspond to the natural numbers. But (a) there is more than one way to define a set of objects that behave like the natural numbers, and (b) it is questionable how "enlightening" such a definition is. In considering larger parts of mathematics, the situation may be somewhat different, since we are there providing definitions of many sorts of things in terms of a single kind of entity: the set.

Often when we introduce a primitive predicate (or operator) into our language (e.g., "red"), we suppose that we can tell, at least in a certain domain, what objects satisfy it. (It is sometimes claimed that we can learn such terms by "ostension" — but more of ostension later.) The predicate "N" in this respect is different from the predicate "red." To be sure, we can tell five golden rings from two turtle doves, but the natural numbers 2 and 5 are abstract objects, and not to be encountered in the world. What role does either the number 5 or the predicate "N" play in our intellectual economy?

One way of approaching this role is to go back to kindergarten: if John has five apples and Mary has two, how many do they have jointly? Seven, because five apples, combined with two apples, are (under the right conditions — sometimes they combine to make applesauce) seven apples. This is true also of oranges, of pieces of fruit, and of distinct individuated objects of any sort. By something akin to what Quine calls "semantic ascent,"[13] we collect the appropriate instances of

---

[13] W. V. O. Quine, *Word and Object*, John Wiley, New York, 1960, p. 271.

⌜5 *A*'s together with 2 *A*'s are 7 A's⌝ into the abstract generalization "2+5=7," which in turn is codified by our theory. If our world were not full of distinct, individuated objects, we would have less use for arithmetic (though arithmetic would be no less true), and if there were none, we would perhaps never have invented arithmetic at all (I guess one *could* invent arithmetic without distinct, individuated marks on paper).

It is sometimes suggested that arithmetic is useful because the world of distinct, individuated objects is a *model* of arithmetic as a formal system (in the semantic sense briefly described previously). This seems doubtful. Suppose that the world were a fluid, and fluid through time, except for a thousand apples and John and Mary. Surely we could use arithmetic to calculate how many apples John and Mary have jointly, given how many each has, even though, due to its finitude, the world of individual objects is *not* a model of arithmetic. The suggestion that the world should constitute a model of arithmetic in order for arithmetic to be useful seems to put the cart before the horse: what matters to us directly is how many apples there are and how many apple pies they will make; arithmetic is handy for knowing such facts, but its handiness does not depend on the world of apples being a model of arithmetic. It suffices that the world of apples we are concerned with be a partial or truncated model of arithmetic.

The notion of truncated model is easy enough to define. If a model of a first-order theory consists of a universe U, together with subsets of U^n to represent n-place predicates, and functions from U^n to U to represent n-place operations, then a *truncated* model consists of a (perhaps finite) subset TU of U, together with subsets of $(TU)^n$ to represent (partial) predicates, together with (partial) functions from $(TU)^n$ to TU. Obviously the axioms of almost any interesting theory will be false in the truncated model. But the distinction between models that can be formally extended to models of the axioms and those that cannot is still worth making.

Is the world even a truncated model of arithmetic? When you combine one thing and another, arithmetic requires that you always get two: but suppose that the things are rabbits? Or drops? Such matters need not hold us up, though; the world is full of things that we *can* discriminate and individuate and that we have every reason to believe comprise truncated models of the natural numbers. It is our ability to make such judgments appropriately that makes the arithmetic of natural numbers useful to us.

In general, where arithmetic matters to us in science, it is because we are performing arithmetic operations on quantities—numbers of apples, numbers of pies, or whatever. For this purpose, we have no need to suppose that there is any complete model of arithmetic in the world, though as we consider different kinds of things, we may require larger numbers. This issue becomes pressing when we consider abstract objects. Even if n is large enough to measure the number of apples in the world, we may still need a larger natural number (e.g., 2^n) to talk about the number of subsets of apples. So despite the fact that there is an upper bound (though we don't generally know it) to how many entities of a given kind there are, it is easy to come up with new kinds of entities of which there are exponentially more.[14]

Geometry

Geometry is curiously different from the kind of mathematics we have been dealing with so far. We can, and do, provide first-order formal systems whose primitive terms correspond to "point," "line," and so on. Since the eighteenth century, a variety of these systems have been studied, in addition to the one, Euclidean geometry, we all studied in high school. These systems were mathematically useful but were thought to have no physical application. Early in this century, however, Einstein made use of a non-Euclidean geometric framework for his successful theory of relativity. Since then, it has been common to say that *now* we know that space is *really* non-Euclidean.

Since we no more encounter points and lines than we encounter those other abstract objects, numbers, we cannot say that we have "discovered" that points and lines are related in non-Euclidean ways. Points and lines, however, are closer to objects of ordinary experience than are pure numbers (as opposed to numbers of A's). We can, for example, take a tiny bundle of light rays and say that it very nearly traces out a line in space. If we talk this way, and if we are talking about intergalactic spaces, and if Einstein's physics is true, then indeed

[14] If it were of intrinsic interest, one could formalize a finite arithmetic consisting of the numbers 1 through 1000, together with the special number Too-Big, by adding the axioms that Too-Big > 1000 and that is $x \geq 1000$, $S(x)=x$, and modifying the axiom that if $S(x)=S(y)$, then $x=y$ to exclude the case where $x=1000$ or $y=1000$, since $S(1000)=$ S(Too-Big) but Too-Big > 1000. Such an arithmetic would work perfectly well for kindergarten problems. It is also interestingly similar to arithmetic in certain primitive cultures in which there is only a small supply of number words, one of which is "many."

the relations among the geometric objects interpreted in terms of light rays are non-Euclidean.

If we talk this way about geometric objects interpreted in terms of light rays near the surface of the earth, or under water, the geometry is non-Euclidean, but not in any nice way. Any simple and sensible geometry is false of actual light rays traveling through a refractive medium such as air.

So the question of why anyone should invent Euclidean geometry is just as serious as the question of why we should employ a non-Euclidean geometry. We use Euclidean geometry all the time. We use it for surveying, for navigation (in both cases representing great circles by straight lines!), for designing machinery. It is often said — even the preternaturally perspicacious Susan Haack says it[15] — that we can *get away with* using Euclidean geometry for these purposes because the distances involved are (in intergalactic terms) relatively short, and over short distances Euclidean geometry provides a close approximation to the *true* non-Euclidean characterization of space.

Against this, it can be argued that the geometry appropriate to navigation, surveying, and machine design is *exactly* Euclidean. Let us proceed from the small to the large. Machine design requires that three flat, rigid surfaces be such that surface A can be in contact with surface B all over, surface B can be in contact with surface C all over, and surface A can be in contact with surface C all over. The flat surface — the plane — of machine space therefore cannot be curved, either positively or negatively: it *must* be Euclidean. Furthermore, we have ways of constructing (approximations to) these flat surfaces. The results are imperfect, to be sure (and "rigidity" has it problems), but the ideal is exactly the Euclidean geometry. We interpret "plane" as the limit approached by such physical constructions, "line" as the intersection of two such planes, and so on. High school geometry is approximately true of such constructions, and the more accurate the constructions, the more nearly true.

The surveyor identifies "line" with his line of sight (and "point" with a peg driven into the ground, but never mind that). But, as in the case of the machinist, the identification is approximate. If there is a mass of hot air causing refraction between the points he is shooting, he will wait for better conditions, but, barring that, he will try to find

---

[15]Susan Haack, *Philosophy of Logics*, Cambridge University Press, Cambridge, 1978, p. 228.

some way of compensating for the distortions introduced by refraction. He does not take the *actual* path of a light beam as a straight line, but instead *corrects* for distortions.

Suppose that we lived on a very heavy planet, with its own heavyweight Einstein. Having discovered that the geometry of light rays was non-Euclidean, the heavyweight surveyor (if there were not a whole new procedure for surveying) would (or could) learn to compensate for the discrepancy between the path of the light ray giving him his shot of a distant object and the Euclidean straight line to that object.

Note that these two examples are not unrelated: when the surveyor draws his map, he employs a *mechanical* (i.e., Euclidean) straight edge. So, to reflect accurately the reality of one scale on a map of another scale, one must *force* the geometries of machine construction and of surveying to be the same.

The navigator, even of a local spacecraft, also uses Euclidean geometry. In this case, however, it is easier to imagine that a navigator near a very heavy star might have to compensate for the "curvature" (in Euclidean terms) of light rays. To compensate thus, of course, is exactly to employ Euclidean geometry as the norm. And he would do so, barring a whole new (computerized) system of navigation, because the *instruments* with which he plots his position on the star charts are Euclidean. The instruments are not Euclidean because they are *small*; they are Euclidean because we built them that way. In fact, this would be no new problem: navigators for hundreds of years have had to plot their positions on the curved surface of the globe with flat maps and Euclidean straight edges.

In each case, the formal ideal, the ideal one approaches as closely as one's instruments allow, is the Euclidean ideal. And nothing precludes an arbitrarily close approximation. Thus the appropriate question when faced with a plurality of geometrical formal systems, is not "What is the true nature of space?" but "What space are we trying to represent? And what are we trying to represent it *for*?" In the case of celestial mechanics, we get one answer; in the case of automobile mechanics, we get another.

3
Probability

Varieties of Probability

Nondemonstrative arguments yield probabilities rather than certainties. Scientific arguments—that is, arguments supporting general empirical hypotheses—can yield probabilities at best. (Many would say that they cannot even yield probabilities.) The uncertainties of the premises, even of a demonstrative argument, may infect the credibility of its conclusion. Such is conventional wisdom concerning the arguments of real life.

Despite the fact that this understanding has been with us for a long time, efforts to provide a *measure* of uncertainty have been made only since about the sixteenth century. This is curious, since the quantitative uncertainties of gambling and of gambling apparatus have been around since the beginning of history and before, and gambling cannot exist without the participants having some measure—if only vague and directly empirical—of the chances involved. But it is not easy to make the connection between the uncertainties of chances and the uncertainties of arguments. The inconclusive work of the past 200 years concerning the foundations of probability and nondemonstrative inference reveals this clearly.

There are two roots to the notion that concerns us, which we will call henceforth simply "probability." One root lies in mass phenomena: the behavior of gambling apparatus initially, social statistics subsequently, and currently phenomena in nearly every domain of knowledge, from genetics to quantum mechanics. The other root lies in the uncertainties that infect the conclusions of arguments. (In fact, according to Hacking,[1] the origin of the use of "probable" in the modern sense lies in theological disputes in which the evidence—the authorities—did not yield a univocal answer.) Courts of law provided the context for some of the considerations developed in this area. Decision theory—based on the principle of maximizing expected utility—had its formal beginning in the work of Pascal.[2] And decision theory is no less pervasive today than the study of mass behavior; it is a subject of research in psychology, business, economics, and political science, as well as in philosophy.

These two roots have given rise to the two main classes of interpretation or definition of probability. Considerations of mass behavior suggest construing the probability that an A is a B as the proportion of A's that are B's. If the A's are infinite in number, "proportion" doesn't make much sense, so it has been suggested (von Mises,[3] Reichenbach[4]) that the probability that an A is a B be defined as the *limit* of the relative frequency of B's in a finite initial segment of the A's as the length of that initial segment is increased without bound. More abstractly, it may be defined as the *measure* of the set B in the set A, where "*measure*" is a purely mathematical notion, satisfying roughly the same constraints as "*proportion*." Probability, in any of these views, is an *empirical* notion characterizing the relation of the frequency or measure of one class or sequence in another. It reflects the *facts* about A and B.[5] This empirical interpretation of probability is alleged to underlie classical statistics[6] and to reflect the main use of probability in genetics, quantum mechanics, and so on.

The other root, according to which probability is a degree of certainty, has proliferated even more. It has given rise to the logical inter-

---

[1] Ian Hacking, *The Emergence of Probability*, Cambridge University Press, Cambridge, 1975. This book, though idiosyncratic, provides an interesting background for the contemporary study of probability.

[2] B. Pascal, first published, I believe, in *The Port Royale Logic*. Arnauld, *The Art of Thinking*, Bobbs-Merrill, Indianapolis, 1964. (Originally published in 1662.)

[3] R. von Mises, *Probability, Statistics and Truth*, Macmillan, New York, 1937. (First published in German in 1928.)

[4] H. Reichenbach, *The Theory of Probability*, University of California Press, Berkeley, 1949. (First published in German in 1934.)

[5] An exception: A and B may be logical or set-theoretical entities, and the "facts" about them may be logical facts.

[6] J. Neyman, one of the founders of standard statistical theory, is quite explicit about this in his *First Course in Probability and Statistics*, Henry Holt, New York, 1950.

pretations of probability (Carnap,[7] Hintikka[8]), various views of "support" both in philosophy and in computer science that do not satisfy the usual axioms for probability (Kemeny and Oppenheim,[9] Hempel and Oppenheim,[10] Popper,[11] Shortliffe[12]), and, recently, *subjective* interpretations of probability according to which the probability of a statement for an individual is reflected in the degree to which an individual is willing to act on it.[13] This last interpretation is neatly tied to the idea of making decisions under uncertainty, and has become popular in business schools as well as in philosophical circles and computer science. Probability, according to any of these views, attaches primarily to statements or to unique events. Its value depends on the evidence (in the case of logical views) or on the opinion of the judging subject (in the case of subjective views). In neither case does it depend *directly* on frequency facts about the world. Probability is either a logical notion or, on the subjective view, an empirical notion of a very different sort than on frequency-related views.

Many proponents of many of these views have argued that their interpretation is the correct (or the most useful, or the only useful) interpretation of probability. Faced with this situation, the reasonable

---

[7]R. Carnap, *The Logical Foundations of Probability*, University of Chicago Press, Chicago, 1950.

[8]J. Hintikka, "A Two-Dimensional Continuum of Inductive Logic," in J. Hintikka and P. Suppes (eds.), *Aspects of Inductive Logic*, North-Holland, Amsterdam, 1966, 113–132.

[9]J. Kemeny and P. Oppenheim, "Degree of Factual Support," *Philosophy of Science* 19, 1952, 307–324.

[10]C. Hempel and P. Oppenheim, "A Definition of 'Degree of Confirmation'," *Philosophy of Science* 12, 1945, 98–115.

[11]K. Popper, "The Propensity Interpretation of the Calculus of Probability and the Quantum Theory," in S. Korner (ed.), *The Colston Papers*, Butterworth's Scientific Publications, London, 1957, 65–70.

[12]E. H. Shortliffe, *Computer-Based Medical Consultations: MYCIN*, American Elsevier, New York, 1976.

[13]This idea goes back to F. P. Ramsey, *The Foundations of Mathematics*, Routledge, Kegan-Paul, London, 1931, and B. de Finetti, "La Prevision: ses lois logique, ses sources subjectives," *Annals de l'Institut Henri Poincaré* (1937), translated in H. Kyburg and H. Smokler (eds.), *Studies in Subjective Probability*, Krieger, Huntington, N.Y., 1980. The best-known philosophical exponent of this point of view is R. C. Jeffrey, *The Logic of Decision* (2nd ed.), University of Chicago Press, Chicago, 1984. L. J. Savage, *The Foundations of Statistics*, John Wiley, New York, 1955, is better known to statisticians.

and judicious Rudolf Carnap proposed[14] that probability was ambiguous and that two interpretations ("explications," as he put it) were called for. Probability$_1$ was to be taken as the concept relating to degree of certainty, the logical force that evidence gives to a conclusion; and probability$_2$ was to be taken as the empirical notion, related to mass behavior and relative frequencies.

Socially judicious this may have been, but it left unexplained the oldest and clearest example of the use of probability: gambling. Gambling apparatus is designed to exhibit a certain sort of mass behavior, and gambling apparatus is designed in order to give people the chance to make decisions on specific, unique outcomes of trials of that apparatus. For the gambler, degrees of certainty and measures of mass behavior somehow come together. How?

Answers come from both camps. The hard-nosed answer from the frequentist camp is that probability just *is* frequency (or measure) and that it is therefore nonsense for the gambler to talk of the probability of heads on a *specific* trial of the coin-tossing apparatus. Nevertheless, the frequentists can say something: they can say that if the gambler does *not* make his odds conform to frequencies or measures, then in the long run he is "practically certain" to come out on the losing end. But this is less helpful than it sounds, since "practical certainty" cannot be construed as high probability on his view: a specific sequence of trials is no more a fit subject for probability assertions than a specific trial.

A softer answer (coming from K. R. Popper,[15] among others) is that in certain (hard-to-specify) kinds of cases, we may take the frequency or measure in a class and distribute it as a *propensity* or *chance* possessed equally by *each member* of that class.

The next roll of the die (if rolls of dice constitute an appropriate case) has the same chance or propensity to come up six as any other, and so that chance itself is what one should use in calculating expected utilities and thus the odds for bets. Sometimes. The *last* roll of the die has (had? But then what happened to our distribution principle?) the same chance, but it is not ordinarily a fit subject for bets. Why not?

The other camp also has an answer to the question of the relation between frequency and decision. It lies in a famous theorem due to de

---

[14]Rudolf Carnap, *The Logical Foundations of Probability*, University of Chicago Press, Chicago, 1950.

[15]K. Popper, op. cit.

Finetti.[16] This is the theorem: suppose that you and I contemplate a sequence of events (tosses of a coin), and the subjective probabilities we assign to these events and combinations of them satisfy some very mild conditions (roughly: we regard any *order* of heads and tails in the same light as any other order, and we don't assign probabilities of 0 or 1 to any logical possibilities). Suppose that we take as evidence an initial segment of this sequence—say, the first n trials. Then our opinions, as reflected by our subjective assignments of the probability of heads on the $n+1$st toss, will converge, as n increases, toward the relative frequency among the first n, and therefore toward each other. This theorem is alleged both to draw the teeth from the charge of excessive subjectivism and to provide a connection between subjective probabilities and observed frequencies.

Neither answer is entirely persuasive. It is not clear why the "practical certainty" invoked by the frequentist cannot be divided up, under some epistemic circumstances, into parts. If we can be practically certain that a very long sequence of tosses will exhibit about 50% heads, why can't we be half practically certain that a single toss will yield heads under just the same epistemic circumstances? The subjectivist points out that for every reasonable diversity of opinion you and I may have, and for every standard δ of agreement, there is an n such that we can be sure after n trials that we agree within δ. It is also true that for every standard δ of agreement and every number of trials n, there is some reasonable initial diversity that you and I may have that does not yield agreement within δ.[17]

Basic Intuitions

Formally, the foundations of probability are complex. The complexity stems from having to deal with a body of knowledge in which inference is nonmonotonic. That is, as Fisher noticed in 1936,[18] the validity of uncertain or probabilistic inference, as distinct from ordinary de-

[16]De Finetti, op. cit.

[17]H. Kyburg, *Epistemology and Inference*, University of Minnesota Press, Minneapolis, 1983, contains informal expositions. More detail is to be found in H. Kyburg, *The Logical Foundations of Statistical Inference*, Reidel, Dordrecht, 1974, and H. Kyburg, *Probability and the Logic of Rational Belief*, Wesleyan University Press, Middletown, Conn., 1961.

[18]R. A. Fisher, "Uncertain Inference," *Proceedings of the American Academy of Arts and Sciences* 71, 1936, 245–258.

ductive inference, can be undermined by the addition of premises—that is, by an increase in the contents of the body of knowledge relative to which the probability of the inference is judged. In the next section we shall try to give some idea of the way in which that complexity can be handled, but here we shall try to explore some of the basic intuitions in a way that is more easily accessible.

Suppose that we are interested in the probability that Suzanne will have been married before her 20th birthday. There are many frequencies that might be relevant: the frequency with which women in general, in the whole world, get married before they reach 20; the frequency with which middle-class American women get married before they reach 20; the frequency with which women named Suzanne get married at all; the frequency with which women who are engaged to be married on a certain date actually get married on that date; and so on. The general principle I want to defend is that all probabilities are based on known frequencies. We do know something about the frequencies already mentioned: the first is very high, the second relatively low, the last quite high. We know the first frequency quite precisely (surely it is between .90 and .99), but we know the others more vaguely (we could put the second between .2 and .5, the last between .7 and 1.0).

Our problem, then, is to provide grounds reflecting all the things, or at least all the *relevant* things, we know about Suzanne and about marriage customs, and so on, for choosing one of these intervals to be the probability of the proposition in question. It is this that is formally very complicated. But intuitively it is just a matter of getting the right reference class, and we have some intuitively clear ideas about that. For example, if we know the frequencies mentioned above, and know of Suzanne only that she is a middle-class American woman, the low frequency associated with middle-class American women is clearly more appropriate than the high frequency associated with women in general. And if we know that in fact Suzanne got married when she was 18, the frequency and probability should both be 1.0.

So a probability is just a frequency (or an interval of frequencies) in a reference class, and our only problem—often trivial—is to select the right reference class.[19]

The ordinary axioms and rules of probability really concern frequencies or their counterparts, and in that respect are trivial tautolo-

---

[19]For a formal specification of how to go about it, see H. Kyburg, "The Reference Class," *Philosophy of Science* **50**, 1983, 374–397. The principles involved will be discussed informally in the next section.

gies (in the finite case) that have nothing especially to do with probability. If a sixth of the tosses give "one" and a sixth of the tosses give "two," then since "one" and "two" are exclusive outcomes, a third must give either "one" or "two." Usually (but perhaps not always) the probabilities of different statements can be referred to a common reference class, and so the probabilities usually (but not always) satisfy the conventional axioms and rules.

In general, the frequencies on which we base our probabilities are empirical frequencies that we have learned hold in the actual world. We either learn these things directly (it is by accumulating statistical evidence that we learned that the frequency with which human births are births of males is about .51) or we infer them from other bits of empirical knowledge (coins are fairly symmetrical physically, and very few people can control a coin toss, so the relative frequency of heads among coin tosses is close to one-half).

Note that a statement whose probability we are interested in may not wear the corresponding frequency on its sleeve. Back to Suzanne: suppose that we know that she will be 20 tomorrow, that she is engaged to be married today, and that, feeling qualms, she has chosen to decide whether to marry or not by tossing a coin and going through with the marriage if and only if the coin lands heads. If we really know all this, then the appropriate reference class for Suzanne's getting married before she is 20 is not any class of women but a class of coin tosses. More precisely: the probability that Suzanne will have been married by the time she reaches 20 is one-half, because this statement will be true if and only if the coin Suzanne tosses lands heads, and the reference class for that particular coin toss is the set of coin tosses in general. So much, also, for the coin that is to be tossed only once and then melted down: the right reference class is not tosses of *that* coin, but tosses of coins in general.

There are also some relative frequencies that can be known a priori. Consider, for example, a very large finite class A. A certain proportion of members of that class also belong to class B. But *whatever* that proportion may be, it can be shown on set theoretical grounds alone that the frequency with which *subsets* of A have approximately the same relative frequency of B's as the whole class A does is high. That is, practically all samples among the set of possible samples are representative. Of course, in a particular circumstance, the set of all possible samples may or may not constitute an appropriate reference class for the probability. But the point here is that bounds on the relative frequency of representativeness in this class are computable a priori.

The upshot is that all probabilities are based on *known* relative frequencies (or counterparts thereof, such as chances) but that they are not to be *identified* with these frequencies. The frequencies may be known a priori on the basis of set theory, directly empirically on the basis of inference from statistical samples, or on the basis of indirect empirical knowledge. (In general, a probability will not be based directly on a frequency in a sample, since in general, the object of interest will not be a member of the sample.) To decide *which* frequency should represent the probability of a particular statement involves deciding which reference class should determine that probability. Given knowledge of frequencies, we still have to choose the right reference class. On the interpretation of probability to be presented in the next section, that is the hardest and most complicated job: providing principles for choosing the right reference class.

Evidential Probability

We will use the following interpretation of probability: it is to be a metalinguistic function from sets of sentences (rational corpora, bodies of knowledge) and sentences to subintervals of the closed interval [0,1]. Thus the domain of the probability function, if we (ill-advisedly!) suppress the reference to the corpus representing evidence, is sentences, and this view must be construed as falling within the degree-of-certainty class of views. But we pay our dues to frequencies by insisting that *every* probability be based on statistical *knowledge*. Furthermore, by interpreting the indefinite article "a" as implying randomness, and by considering a (hypothetical) corpus containing true and exact statistical or measure statements about the world, we get a precise metalinguistic reflection of the frequency interpretation.

What we fail to get is an interpretation that satisfies the axioms of the probability calculus in all cases. But that is to be expected, since the probability calculus concerns real-valued measures, and evidential probabilities are interval valued. As we shall see, however, there are interesting and useful relations between these intervals and ordinary probability measures.

How does the definition of probability go? Suppose that K is a rational corpus consisting of sentences in our first-order language L, and that S is a sentence of L. Then the probability of S relative to K is the closed interval $[p,q]$ — symbolically, $Prob_L(S,K)=[p,q]$ — just in case:

1. S is known in K to be equivalent (in truth value only) to a statement of the form ⌜$x \in y$⌝; that is, ⌜$S \equiv x \in y$⌝ belongs to K.
2. The rational corpus K contains knowledge of the frequency (or measure) of y in z. We express this knowledge by the sentence ⌜$\%(y,z) \in [p,q]$⌝; thus we are requiring that this sentence belong to K.[20]
3. The sentence ⌜$x \in z$⌝ belongs to K; furthermore, x is to be a *random member* of z with respect to y, relative to K.

This establishes the connection between statistics known in K and probabilities, but it leaves us with the problem of explicating (3); how are we to construe the phrase "random member of?"

Randomness, like probability itself, will be construed as a metalinguistic, epistemological relation. It is a relation between K and the three terms x, y, and z. Clearly we require that K contain the sentence ⌜$x \in z$⌝; x can hardly be a random member of z with respect to anything, relative to K, unless it is known in K that x is a member of z. But what else do we require?

The easiest way to get at these further requirements is to consider competing *inference structures*.[21] An inference structure may be construed as a set of candidates x, y, z, $[p,q]$ for determining the probability of S according to our definition. An inference structure thus has the form[22]

$$S \quad x \quad y \quad z \quad [p,q]$$

Inference structures for our sentence S will all begin with S; they will all be such that ⌜$x \in y$⌝ is known to be truth functionally equivalent to S (i.e., in K); and they will all be such that ⌜$\%(y,z) \in [p,q]$⌝ is in K for some candidate interval $[p,q]$. (If ⌜$\%(y,z) \in [p,q]$⌝ is in K, and the

---

[20]One way of articulating ⌜$\%(y,z) \in [p,q]$⌝ would be "the *statistical* probability that a z is a y lies between p and q." But "statistical probability" — if it is not merely an anomalous locution! — is a proportion or relative frequency or measure, and to assert that the probability that "a" z is a y is in $[p,q]$ is exactly to say that it is *known* that the proportion or measure of y's among z's is in that interval. We have no need of a separate notion of probability, though it is unlikely that it can be eradicated.

[21]This relatively new and simple approach is laid out in detail in "The Reference Class," loc. cit.

[22]For serious statistical purposes, it is often advantageous to replace y by an ordered pair $<w,b>$, where w is a quantity and b is a Borel set, and what is known is that S has the same truth value as ⌜$w(x) \in b$⌝.

consequences of each sentence in K are in K, then $\ulcorner(y,z)\in[p-\epsilon, q+\delta]\urcorner$ is in K. What we are interested in, however, is the *strongest* statistical statement about z and y in K. We assume henceforth, when we write $\ulcorner\%(y,z)\in[p,q]\urcorner$, that there is no positive ϵ or δ such that $\ulcorner\%(y,z)\in[p-\epsilon, q+\delta]\urcorner\in K$).

Among these structures, some will disagree with others in the following technical sense: We say that

$$S \quad x \quad y \quad z \quad [p,q]$$

disagrees with

$$S \quad x' \quad y' \quad z' \quad [p',q']$$

just in case neither interval mentioned is a subinterval (proper or improper) of the other.

When two inference structures disagree, one may *dominate* the other. This can happen in three ways. In the simplest case, x is the same term as x', y is the same term as y', and it is known that z is a subset of z' — which is to say, the sentence $\ulcorner z\subset z'\urcorner$ is a sentence in the set K that represents the body of knowledge with which we are concerned. Thus if we know that Opus is a penguin bird, then our knowledge about the flying abilities of penguin birds dominates our knowledge about the flying abilities of birds in general when we are concerned with the chances that Opus can fly.

Since we don't care about renaming, we can say more generally that of two inference structures that disagree,

$$I = S \quad x \quad y \quad z \quad [p,q]$$
$$I' = S \quad x' \quad y' \quad z' \quad [p',q']$$

the first dominates the second whenever there is a third inference structure of the form

$$I'' = S \quad x'' \quad y'' \quad z'' \quad [p,q]$$

such that the sentence $\ulcorner z''\subset z'\urcorner$ belongs to K. We say that the first inference structure is *reflected in* the second.

A second sort of dominance (to which the first is in fact reducible) arises when one of the inference structures takes account of more statistical evidence than the other. For example, suppose we have 50 black balls and 50 white balls distributed in three urns, with 20 black balls in one urn and 15 black balls and 25 white balls in each of the others. Let a be "the next ball to be drawn" and b "the next urn chosen." Suppose that we know that a is drawn from b; y is the set of black balls, z is the entire set of balls, y' is the set of pairs of objects of which the second is a black ball, and z' is the set of pairs $<v', w'>$ such that v' is a member of the set v of urns and w' is a ball in the urn v'. S, of course, is the sentence that says that a is black. The disagreeing inference structures are

$$I_1 = S \quad <b,a>y' \quad z' \quad [0.583, 0.583]$$

since $1/3 \times 1 + 2/3 \times 15/40 = 0.583$

$$I_2 = S \quad x \quad y \quad z \quad [0.5, 0.5]$$

Again, we would like the first inference structure to dominate the second. The condition that characterizes this sort of case is that there exists an inference structure of the form

$$I'' = S \quad <x,a>y'' \quad z'' \times z \quad [0.5, 0.5]$$

matching the second, whose reference set is the cross-product of the second reference set with something else, and that there is another inference structure,

$$I = <x,a>y'' \quad z^* \quad [0.583, 0.583]$$

matching the one we desire, where we know that z^* is included in $z'' \times z$. In the example, we take x to be b, y'' to be y', and z'' to be v. Again, we say that there is a reflection of the first inference structure in the second, or that the second *reflects* the first.

As the example should suggest, this form of dominance is the one involved when we have the statistical background knowledge to employ Bayesian conditionalization. But note that it applies only when the inference structures *differ* in the technical sense. If our prior information is vague, the corresponding inference structures needn't differ, and dominance is irrelevant.

The third form of dominance is that in which the second term (usually denoting a sample) of one inference structure is a subset of (or corresponds to a subset of) the second term (denoting a larger sample) of the other. This case arises primarily in statistical inference. Suppose the two inference structures are

$$I = S \quad x \quad y \quad [p,q]$$
$$I' = S \quad x' \quad y' \quad [p',q']$$

where x is a sample of size 2000, x' is a sample of size 1000, and z and z' are the sets of all samples of the corresponding cardinalities. To fix our intuitions, suppose that x is a sample of 2000 from an unknown population, and that half of the members of the sample have the property P. Take y to be the set of 2000-member samples that have approximately the same proportion of P's there are in the unknown population. Practically all members of z are members of y, so that $[p,q]$ is an interval close to 1. Let x' be a subsample of x that contains just the items with the property P, and take y' to be defined so that ⌜$x \in y \equiv x' \in y'$⌝ is in K. Now if x is representative, we can be quite sure that x' is not; $[p',q']$ is an interval near 0. Clearly we want the inference based on the 2000-member sample to dominate that based on the 1000-member sample.

We say that the first inference structure I dominates the second inference structure I' whenever there is an inference structure

$$I^* \quad S \quad x'' \quad y'' \quad z'' \quad [p,q]$$

matching I and an inference structure

$$I'^* \quad S \quad x''' \quad y''' \quad z''' \quad [p',q']$$

matching I' such that it is known that x''' is included in x'' but not vice versa.

This is no doubt a rather cryptic characterization of dominance, but it may suffice to give the flavor of the notion. More details can be found in "The Reference Class," already mentioned.

Let us now return to randomness. We have defined disagreement and dominance, given K, as syntactic relations between inference structures. We have left to one side the case where $[p,q]$ in one inference structure is a proper subinterval of $[p',q']$ in the other inference structure. In this case, we say that the first inference structure is *stronger than* the second. And now we can define randomness quite simply: we say that x is a random member of z with respect to y, relative to the corpus K, just in case the inference structure

$$S \quad x \quad y \quad z \quad [p,q]$$

is one of the strongest inference structures for S that dominates every inference structure for S with which it disagrees ("one of" since, in general, there will be a number of distinct, equally strong inference structures satisfying this constraint).

Relatively detailed and formal expositions of this conception of probability are to be found in the works mentioned above. All of those expositions contain more detailed justification of the three kinds of dominance, and a number of examples. The object here was simply to present enough of the underlying ideas to give the reader a feel for the concept of epistemological probability.

Properties of Probability

Let us look at the properties of the metalinguistic, syntactical, evidential notion of probability just characterized. Since the values of the probability function defined are intervals, we do not have the classical calculus of probability. But, depending on K, we can derive some probability relations. We ordinarily suppose K to contain some form of set theory, or whatever mathematics we need. In addition, I shall suppose that K is weakly deductively closed. By this, I mean that if a statement S is in K, and S logically implies T, then T will also be in K. It is often supposed that a body of knowledge should be strongly deductively closed, containing the logical consequences of any set of statements it contains or, what is the same thing in the presence of weak deductive closure, containing the conjunction of any two statements that belong to it. I shall not assume this generally, for reasons that will become clear later.

Alternatively, it is sometimes argued that in view of human finitude, even weak deductive closure—including all the consequences of every *single* statement in the corpus of knowledge—is too strong a constraint. Since every logical theorem is a consequence of any single statement, no one can satisfy this condition of weak deductive closure.

But the rational corpus—the formal syntactical object—should be looked on as an ideal to be approached rather than as a state that a rational being (person or computer) should be in. An ideal need not be achievable to function constructively as a goal. But it is necessary that we be able to tell whether or not we are closer to the ideal at one time than at another, and it is necessary that we be able to approach the ideal. We can use weak deductive closure this way: providing a proof of conclusion C from the single premise $\ulcorner P_1 \& P_2 \& \ldots P_n \urcorner$ shows—if you rationally accept the premises—that you will be closer to the ideal by also accepting the conclusion. This principle becomes far less plausible if you allow the arbitrary accumulation of premises; at some point, their conjunction becomes so incredible that they no longer provide good grounds for accepting their implications. Thus we do not go so far as to adopt full deductive closure.

We shall also assume that K is consistent in the very weak sense that it contains no explicit contradiction—that is, no statement of the form ⌜$S \& \sim S$⌝. Given that K is weakly deductively closed, and weakly consistent, we can establish the following properties:

1. If S is known to have the same truth value as T—that is, if ⌜$S \equiv T$⌝ is in K—then $Prob(T,K) = Prob(S,K)$.

2. If $Prob(S,K) = [p,q]$, then $Prob(\sim S, K) = [1-q, 1-p]$.

3. If $Prob(S,K) = [p,q]$ and $Prob(T,K) = [r,s]$ and K contains the statement ⌜$\sim (S \& T)$⌝, and K contains the conjunction of the two statistical statements on which $Prob(S,K)$ and $Prob(T,K)$ are based, then $Prob($⌜$S \vee T$⌝$)$ is included in $[p+r, q+s]$, or vice versa.[23] This yields the existence of an additive measure satisfying the interval constraints.

4. If $Prob(S,K) = [p,q]$ and $Prob(T,K) = [r,s]$, and S implies T (i.e., T is derivable from S), then $r > p$.

5. Let a maximal, strongly consistent subset of K be called a *strand*. If S belongs to every strand of K, then $Prob(S,K) = [1,1]$.

6. If K is strongly consistent and S belongs to K, then $Prob(S,K) = [1,1]$.

A subjective or logical probability function does satisfy the classical calculus of probability. These probability functions are real-valued functions **B** such that for any statement S in a given language L, $B(S) \in [0,1]$, and if ⌜$S \& T$⌝ is logically false, $\mathbf{B}($⌜$S \vee T$⌝$) = \mathbf{B}(S) + \mathbf{B}(T)$. It is possible to show that if K is weakly deductively closed and weakly consistent, then for any finite set of statements in L there exists a function **B** that satisfies the classical probability calculus and, furthermore, has the property that for any statement S in this set whose probability exists, $\mathbf{B}(S) \in Prob(S,K)$.

For special kinds of K, we can get something corresponding to the classical calculus. Suppose that K consists of the logical consequences of a statement saying that the random quantity Q is distributed in the reference class R according to the distribution function F, together with the statement "$a \in R$." Let B be a Borel set—that is, a set of real numbers obtainable from intervals, complements of intervals, and denumerable unions of such sets. Then the probability of any state-

---

[23]It is not unusual for the probability of a disjunction to be more precisely known than the probabilities of the disjuncts: consider a die known to be biased a bit toward the 1 at the expense of the 2, or vice versa. The probability of a 1 *or* a 2 might be exactly ⅓, though we might have to be a bit vague about either disjunct alone. The opposite: the case in which we are quite sure about each disjunct, but vague about the disjunction, is rather pathological but possible.

ment of the form "$Q(a) \in B$" is the degenerate interval corresponding to the integral of dF over B, and these probabilities (considered as real numbers rather than degenerate intervals) will satisfy the classical axioms.

If we construe the indefinite article "a" in the appropriate way, we can even represent the classical frequency or measure theoretic interpretation of probability. According to this interpretation, the probability that an R is a B is p, just in case the frequency, limiting frequency, or measure of B in R is p. Let us take the indefinite article "a" to mean "every random," where *random* is construed in the epistemic sense. Then (escalating to the metalanguage) "the probability that *an R* is a *B* is p" becomes: for every term t, if t is a random member of R with respect to B, relative to K, $Prob(\ulcorner t \in B \urcorner, K) = [p,p]$. This will be true, according to the definition of the preceding section, if and only if "%$(B,R) \in [p,p]$" is in K.

Now let us take K to be the (deductively closed) corpus of a statistically omniscient deity. That is, we take K to include (whatever else it may include) all and only true statistical statements. By an application of Tarski's lemma, then, "%$(B,C) \in [p,q]$" is in K just in case %$(B,C) \in [p,q]$. In short, if t is "a" member of C, then $Prob(\ulcorner t \in B \urcorner, K) = [p,p]$ if and only if %$(B,C) = p$.

Probability has been defined syntactically here. The assertion "$Prob(S,K) = [p,q]$" is true on syntactical grounds alone, if it is true at all, just as the (metalinguistic) assertion "$S \vdash T$" is true on syntactical grounds, if it is true at all. The probability function is thus a *logical* function in just the same sense as the provability relation is a logical relation. I refrain from referring to the interpretation I have offered as a logical interpretation of probability (I have not always done so) only because it is likely then to be confused with a logical *measure* interpretation such as Carnap's or Hintikka's. Nevertheless, although probability is construed logically, it is important to remember that every statement of probability is based on a (hypothetically or actually) known frequency or measure. That is why, with the help of the statistically omniscient deity, we can recapture the empirical conception of probability.

All probabilities, as we have construed them, are conditional on a rational corpus K. "Conditional probability" is ordinarily construed in quite a different way: given a probability function P defined in sentences, the *conditional probability of A given B*, $P(A|B)$, is just the ratio $P(A\&B)/P(B)$. The famous principle of Bayesian conditionalization is the principle that if you start with a probability function P, and then learn something new that is captured by the sentence B, you should shift to a new probability function P^* defined by $P^*(A) = P(A|B)$.

Since Bayesian conditionalization has played such a large role in the philosophy of science, I should mention how these probabilities are related to conditional probabilities. Let $\mathcal{A}$ be the smallest Lindenbaum-Tarski algebra of L containing A and B, and let $\mathcal{P}$ be the set of all classical probability functions P defined on $\mathcal{A}$ satisfying $P(X) \in Prob(X, K)$ for all X in $\mathcal{A}$. By an earlier remark, $\mathcal{P}$ is not empty. Define:

$$Prob(A|B,K) = [\min_{\mathcal{P}} P(A|B), \max_{\mathcal{P}} P(A|B)]$$

We may also consider what happens when the statement B is added to our body of knowledge K. Let KB be the union of K and the logical consequences of B. Then another sense of conditional probability is just $Prob(A, KB)$.

It turns out that while it is often true that $Prob(A, KB) = Prob(A|B, K)$, this identity does not hold in general. Examples have been offered by Levi,[24] but a very simple example can be constructed involving independence—a notion worthy of consideration in its own right.

It is usual to say that the statements A and B are probabilistically independent just in case the probability of A given B is the same as the probability of A. This is obviously and importantly true of *stochastic* independence. This is the relation that holds for good gambling apparatus: the chance or frequency of heads, following the occurrence of a tail, is the same as the general chance or frequency of heads.

The relation of stochastic independence is easily seen to be symmetric: heads is independent of tails if and only if tails is independent of heads. This symmetry carries over to probabilistic independence when we adopt the standard definition of conditional probability. But this does not seem entirely plausible: since the probability of heads on the next toss, given that the coin is two-headed, is not the same as the probability of heads in general, the probability that the coin is two-headed, given that the coin landed heads once, is not the same as the

---

[24]Isaac Levi discusses this issue in Chapter 16 of *The Enterprise of Knowledge*, MIT Press, Cambridge, 1980.

probability of two-headedness in general. In general, coin tosses turn out not to be probabilistically independent, and that is contrary to usage.

There is a natural answer to this argument, namely, that it is a quantitative illusion: the influence in one direction (from general frequency to individual outcome) is very great; the influence in the other direction (from outcome to general frequency) is very small. Whether there are adequate grounds for invoking this argument is another question; the weight of tradition, by itself, does not constitute such grounds. Intuitively, knowledge of the long-run relative frequency of heads is relevant to the probability of heads on the next toss; getting heads on the next toss is not intuitively relevant to the long-run frequency.

The important question is whether abandoning the symmetry of relevance provides some conceptual or computational advantages in the formalization and representation of our scientific knowledge. The remaining chapters of this book collectively constitute an argument in favor of a nontraditional treatment of relevance.

Statistics

We have stipulated that *all* evidential probabilities must be based on *known* statements of (approximate) frequency or measure. This is all very well so long as we are considering, say, the corpus of the statistically omniscient deity. But how does it help *us*? Given that we have some statistical knowledge, we may be all right, but can we start out with *no* knowledge of empirical frequencies and find probabilistic justification of such knowledge?

The answer is "yes." This is important for two reasons. One is a philosophical matter: it has been claimed that statistical hypotheses can only become probable relative to corpora of knowledge that already embody substantive assumptions.[25] To call these "assumptions" is to imply both that they could rationally be replaced by alternatives and that the conclusions based on them can be rationally rejected, for example, by replacing or rejecting the assumptions. To call them "presuppositions" is even worse; when I say that our differences stem from

---

[25]See, for example, A. W. Burks, "The Presupposition Theory of Induction," *Philosophy of Science* **20**, 1953, 177-197.

different presuppositions, I have insulated my beliefs from your arguments.

The other reason that it is important to be able to start from the zero point is more practical. Leaving whole bodies of scientific knowledge to one side, forgetting about the general philosophical theme of empiricism, we want to be able to handle a specific subject matter from scratch. While in general we can find some relevant knowledge with which to approach any subject, it would be unfortunate to suppose that there is a set of assumptions concerning that particular subject that could not be questioned.

It is true that we can try to have our cake and eat it too: we can say that every substantive conclusion requires substantive assumptions, and also that every set of assumptions can, in turn, be questioned.[26] For every question we may be able to find an answer. But the possibility of an answer to every question is not completely reassuring when you realize that to every answer a new question can be raised in a systematic way. This precludes the assurance of closure.

To ensure that all this is possible, we must have two things: we must be sure that, however empty our rational corpus may be, there will be some interesting statistical knowledge in it on which to base evidential probabilities. And we must adopt a principle of probabilistic inference or inductive inference that will allow us to accept a statement S in a body of knowledge, given that (in some sense to be made precise) S is probable enough. The latter question deserves a chapter of its own: the next. The former is our present concern.

If our rational corpus K contains some set theory and is weakly deductively closed, there are statistical statements in it. For example, if we define (in the object language) $\%(A,B)$ to be 0 when B is empty or infinite or not a set at all, we will always have in our rational corpus $\ulcorner \%(A,B) \in [0,1] \urcorner$. This isn't very interesting, but we can do better.

Suppose that A is a set that is neither empty nor infinite. Consider A^n, the set of sequences of n (not necessarily distinct) members of A. Suppose, hypothetically, that the proportion of A's that are B's is p. It follows that the proportion of members of A^n that contains k B's is $(k,n)p^k(1-p)^{1-k}$.[27] This gives us the distribution of the proportion of B's in *members* of A^n. The distribution has a mean of p (the average pro-

---

[26] This is the approach of Isaac Levi, op. cit.

[27] "(k,n)" is short for $\dfrac{n!}{k!(n-k)!}$

portion of B's in the A^n's is p) and a variance of $p(1-p)/n$. Clearly, $p(1-p)/n$ is at most $1/(4n)$.

Tchebycheff's inequality (a purely mathematical result) then tells us that the frequency among A^n of sequences in which the proportion of B's differs from p by less than $k/(2n^{1/2})$ is *at least* $1-1/k^2$. Denote the set of such usefully representative sequences by $R(k)$. Then, given that we know that A is a nonempty, finite set, we have a whole useful family of statistical statements in our body of knowledge: all statements of the form:

$$\%(R(k),A^n)\in[1-(1/k^2),1]$$

Can we use these statements for statistical inference? Suppose that we have observed a sequence of n A's and have noted that m of them are B's. Denoting the sample by s, we have "$s\in A^n$" in our corpus K and "$\%(B,A)=m/n$" in K. Taking k, for example, to be 2, we have the inference structure:

"$\%(B,A)\in m/n \pm 1/n^{1/2}$" s A^n $R(2)$ [.75,1.0]

Note that "$\%(B,s)=m/n$" entails the equivalence of "$s\in R(s)$" and "$\%(B,A)\in m/n \pm 1/n^{1/2}$." The question is whether or not s is a *random* member of A^n.

Let us consider inference structures that disagree with the one displayed. There will no doubt be some. For example, if we have observed n A's of which m were B's, we have also observed m A's, all of which were B's, and by a similar construction, this will give rise to an inference structure disagreeing with the one we are considering. But the inference structure concerning the whole sample will dominate the inference structure concerning the subsample, according to our definition.

We might already know the proportion of A's that are B's: some statement "$\%(B,A)=p$" might already be in K. In that case, the conflicting inference structure may have the form

"$\%(B,A)\in m/n \pm 1/n^{1/2}$" s $A^n(m/n)$ $R(2)$ [0,0]

taking p to fall outside $m/n \pm 1/n^{1/2}$ and letting "$A^n(m/n)$" denote the set of n-sequences of A's containing exactly m B's. In this case, the new inference structure will dominate the old. There are many ways, depending on what is in K, in which an inference structure can disagree with our original one and not be dominated by it. But it is clearly possible, especially if *K* is rather sparse, that our original inference structure will dominate every inference structure with which it disagrees.

Suppose that this is the case. Then whatever be the correct inference structure for "%$(B,A) \in m/n \pm 1/n^{1/2}$", it will yield a probability interval that is a subinterval of [.75,1.0]. In any event, then, we can say that the lower bound of the probability is *at least* .75 that the proportion of *A*'s that are *B*'s lies in the stated interval. (In fact, there are stronger inference structures, based on closer approximations than that provided by Tchebycheff's inequality.)

We have considered only a very simple case and have obtained only an approximate solution to it, but it should suffice to show that in the sense of probability offered here, statistical evidence can render a general statistical hypothesis probable, relative to a body of knowledge *K* that itself contains no general statistical knowledge other than that provided by pure mathematics.

In more realistic cases, and in ones more familiar in scientific practice, the body of knowledge *K* with which we are concerned is not so austere. It is often said that in doing statistical inference, one must have some *model* in mind. For example, the model may be a normal distribution with an unknown mean and variance, a Poisson distribution, and so on. Although this is a conventional way of talking, and correctly serves as a reminder that one is not ordinarily starting from scratch, it conceals an important point: if our inferential procedure is to give us results we have reason to believe, it had better be based on a "model" that we have reason to believe applies in the case at hand. Somehow "model" (like "presupposition") seems to allow a degree of arbitrariness that is inappropriate in scientific inference: "we don't *really* disagree, because you and I are using different models." This suggests that results can rationally depend on mere opinion (one man's model being as good as another's) rather than on evidence and argument.

There is a way of achieving the same end that puts the cards on the table more clearly. Rather than speaking of models, it seems preferable to speak of sets of distributions (though the two ways of speaking are roughly equivalent). Thus, rather than speaking of using the normal distribution as a model, we could speak of accepting in *K* the general statistical assertion that the quantity in question has a distribution that belongs to the general set of normal distributions. This is quite analogous to accepting in *K* the assertion that the proportion of *A*'s that are

B's lies in a certain interval, and that suffices perfectly well as a basis for probability statements.

Speaking of sets of distributions has three benefits in addition to calling attention to the fact that using a certain model requires justification. First, it allows for vacuous models, as in our initial example. It is a set-theoretical truth that any nonempty, finite set contains some specific proportion of members of any other set. The "model" is represented by the set of all possible proportions. Second, and perhaps more importantly, speaking of sets of distributions allows us to constrain our models in any warranted way: thus the set of normal distributions for the quantity Q that have a variance of 4 and a mean either between 35.4 and 37.2 or between 45.1 and 46.7 is a perfectly useful set of distributions to put into a statistical inference, but it seems like an overdetermined "model".

The final advantage of speaking of sets of distributions rather than models is that it leads us to make explicit the approximations involved. No quantity, for example, can have a normal distribution in a finite population, and none of the populations we encounter in the world are actually infinite. Thus we can't really mean it when we say of a certain quantity that it is distributed normally with an unknown mean and unit variance. What we *can* mean is that the distribution is close enough to normal with unit variance for all practical purposes.[28] And we can spell this out as precisely as we wish.

Employing this idea, let us consider a final example. Suppose that it is known in K that the quantity Q has a normal distribution with a variance of 1 and an unknown mean. More realistically, of course, what is "known" in K is that Q has a distribution that is *approximately* normal and a variance that is *approximately* 1. Furthermore, the mean is not totally unknown: we might know that it could not be negative, for example, and that it is less than 10^5. Let Q' be a new random quantity, defined on the same class, whose value for an object x is $Q(x) - m$, where m is the unknown mean. From what we know about Q', it follows, to the same degree of approximation, that Q' has a normal distribution with mean 0 and variance 1. Let us add to K the

---

[28]It is, for example, quite impractical to suppose that there is any proportion in any ordinary real-world population corresponding to the set of objects 20 standard deviations removed from the mean of a particular normally distributed quantity. Differentiating among probabilities larger than the confidence with which the statistical statement on which they are based is held is idle; and similarly for "smaller" and 1 minus that confidence.

observation report "$Q(a)=7.0$." We know, from our knowledge of Q', that the relative frequency (or chance) is .99 that Q' will take a value between -2.5 and $+2.5$.

If, relative to K, a is a random member of the class in question with respect to being one of those objects with a Q' value between -2.5 and $+2.5$, then, relative to K, the probability is .99 that a has a Q' value in this range. But in K, to say that a has a Q' value in this range is *equivalent* to (has the same truth value as) saying that m lies in the interval [4.5,9.5].

More detailed treatment of statistical inference here would lead us astray from our main concern. The general point is that our body of knowledge K need contain no general empirical statistical assertions in order for us to be able to assign probabilities, and even large probabilities, to general statistical hypotheses. And if our body of knowledge K does contain general statistical assertions — even in the form of sets of distributions — then we can use that knowledge to render more specific statistical hypotheses probable. Both of these claims are controversial, and we should not pretend that the brief argument given here has settled them.

Classical statistics, for example, which generally adopts a frequency interpretation of probability, denies that statistical evidence can render a statistical hypothesis probable, whatever may be in K. On that view, the statistical hypothesis is either true or false and does not take on truth values in any sequence (of what?) that would make sense of assigning it a probability. On the epistemological interpretation of probability, it does make sense.

Subjectivistic Bayesian statistics has no problem assigning probabilities to hypotheses, but it requires the assignment of exact probabilities, even in the absence of any evidence, and tends to disallow acceptance.

What we need for future reference in this work is only the general form of the claim: statistical properties of classes can be discovered approximately, and with high probability, on the basis of the statistical evidence provided by samples. As a generality, this is just common sense. But since it is disputed by some writers, it was important to provide some argument for its truth. In order to settle what sorts of hypotheses are supported to what degree by what evidence, in the presence of what bodies of knowledge, we would have to provide much more detail. For our purposes, the general idea of representativeness and the general conditions for epistemological randomness will suffice.

4
Induction

Inductive Logic

In the good old days, logic was divided into two parts, deductive and inductive. Deductive logic was the study of the syllogism, and inductive logic was, rather vaguely, everything else. Deductive logic has been enriched to include not only first-order logic but various deviant logics (which we are leaving aside here) and sometimes even set theory. Inductive logic has largely disappeared from the logic curriculum, but now appears in various guises in courses in the philosophy of science and epistemology and, recently, in cognitive science and artificial intelligence.

Construed most generally, inductive logic covers all kinds of nondemonstrative inference. For our purposes, it will be useful to make some distinctions. In its narrowest sense, "induction" is the extrapolation of past relative frequencies into the future. Put otherwise, it is the inference from a sample to a population. I shall call this "statistical induction" and place under that heading all kinds of statistical inference, including that which depends on having some statistical knowledge to start with in the body of knowledge K on which our induction is based.

It might be thought, and it has been argued, that inductive *generalization*—arguments of the form "All observed A's have been B's; therefore (probably) all A's are B's—is merely a special case of statistical induction. What's special about the frequencies 0 and 1? Well, there is something special about them. The conclusion of an inductive generalization is usually expressed in the form "All [or no] A's are B's," and this is, in the case of sets of A's that are infinite, distinguishable from "100% of the A's are B's." More important, though, what is rendered probable by statistical inference from a sample of A's is only that the proportion of B's lies in some interval about the observed sample proportion. If we have observed a ratio of r, what is to be concluded is that the true general ratio lies in some interval $[p,q]$, where $p \leq r \leq q$. Applied to an observed ratio of 1, this yields an interval $[1-\epsilon, 1.0]$ that must be distinguished from the universal generalization even in the finite case. (It can be true even when there are some A's that aren't B's.) We shall thus regard inductive generalization as a distinct form of inductive inference. For the moment, we leave open the question of whether there is any legitimate form of inductive generalization.

In scientific inquiry, most of our attention is focused on laws and theories. Achieving reliable laws and theories is also considered induction. Thus it is alleged that by accumulating evidence, we may discover a law that relates gravitational force, mass, and distance. By putting a lot of evidence together, we may be supposed to arrive at a theory of celestial mechanics. A distinction might be made — we will consider its plausibility later — between discovering *laws* that relate quantities that are already part of our scientific vocabulary, and developing *theories* that involve the introduction of new vocabulary. To leave our options open, let us refer to the former as *nomic induction*, and the latter as *theoretical induction*.

There is one more form of nondemonstrative inference that will concern us. It is not often considered a variety of induction, since "induction" is generally thought of as argument from the particular to the general (in contrast to deductive argument from the general — all men are mortal — to the particular — this man is mortal). This is the nondemonstrative and uncertain inference from "Practically all A's are B's" to "This A is a B." More precisely, let us represent it as the inference from a statistical generalization to a conclusion concerning an *instance* of that generalization. We shall call it *instantial induction*.

We have, then:

    statistical induction
    universal induction
    nomic induction
    theoretical induction
    instantial induction

There is another distinction to be made, orthogonal to the first set of distinctions, that is central to our concerns. That is the classification of inductive arguments according to the character of their conclusions. Barring "demonstrative induction" — which can hardly be considered a form of *non*demonstrative inference — everybody who believes in induction at all supposes that it is characterized by uncertainty. Whether this uncertainty is to be construed as measured by probability or not is controversial. Let us neutrally denote the degree of certainty, plausibility, probability, or whatever it is, by "R." (We need not even suppose that R is a numerical measure.)

The schema for an inductive argument can then take either of two forms (as pointed out by Carl Hempel[1]):

Premises
―――――――
Conclusion with hedge R

or

Premises
――――――― R, characterizing the *inference*
Conclusion

In other words, we may ask whether an inductive argument has the form: from this evidence, it follows that the hypothesis H is probable (plausible, supported, . . .) or the form: from this evidence H follows with probability, plausibility, support . . . so and so? Or perhaps sometimes one and sometimes the other?

Hempel argues persuasively that the latter is the form we should adhere to — that is, that our body of scientific knowledge consists of categorical empirical statements rather than modally modified ones. The evidence we have supports (to some degree or other) the conclusion that the relative frequency of male births among humans *is* close to 0.51, not that it is *probably* close to 0.51.

Thus we include as categorical statements concerning proportions or frequencies or measures ("$\%(B,A)\in[p,q]$" and "the distribution of the quantity Q in the class C is given by F" are construed as categorical, even though they are sometimes expressed with the help of the word "probability") and do *not* include statements involving "probably," "likely," and the like in their epistemic usage. Thus the sort of

[1] C. G. Hempel, *Aspects of Scientific Explanation*, The Free Press, New York, 1965.

statement we are interested in including in our corpus of scientific knowledge, and for which we might like to provide an inductive justification, is "All crows are black" or "The law relating the quantities X, Y, and Z is $f(X,Y,Z)$," and not "All crows are black is probable" or "It is likely that the law relating the quantities X, Y, and Z is $f(X,Y,Z)$."

There are several reasons for choosing to treat scientific knowledge in this way, despite the contrary view of eminent philosophers. Carnap, for example,[2] argues that *acceptance* just represents an approximate and loose way of assigning a high probability to a statement. And there is no doubt that we can exclusively use probability statements as a basis for decision making. But there are also good grounds for adopting an acceptance model of scientific belief.

First, it reflects not only the way we do talk — it is often the case that we talk about scientific hypotheses being "established"; our textbooks distinguish between scientific speculation (explicitly qualified with "likely," "probable," etc.) and what is taken to be scientific fact.

Second, most treatments of scientific knowledge take something as evidence, and that, at least, is accepted. It is possible to avoid this, and to consider a body of knowledge to be a complete algebra of statements over which a shifting probability measure is defined. This probability measure shifts in response to changing evidential probabilities. But this is a very complex picture and is certainly not the way we *appear* to operate.

Third, in the practice of science and engineering, we distinguish between statements whose probabilities we take account of and statements that we "accept as evidence," to use Levi's happy phrase.[3] In designing a bridge, for example, we take it as a fact that the modulus of elasticity of the steel we use lies in a certain range, though in measuring that elasticity, we adopt a procedure that will be wrong one time out of a hundred, say. The interplay between acceptance and probability is complex, but it seems natural to consider acceptance as well as probability.

Finally, we may object to the probabilistic treatment that there is no objective source for the probabilities. If there is no agreed-upon source for the probabilities, it is unclear what their probative force is in decision making. And lacking a systematic and objective underpin-

---

[2]R. Carnap, "On Rules of Acceptance," in I. Lakatos (ed.), *The Problem of Inductive Logic*, North-Holland, Amsterdam, 1968, pp. 146-150.

[3]Isaac Levi, *Gambling with Truth*, Knopf, New York, 1967, pp. 28, 32.

ning for probabilities, it is not clear how we can avoid the explosive consumption of our computational resources, whether they are social, in the community of scientists, psychological, in the individual, or artificial, in the computer model of the scientific process.

If we are comfortable with subjectivity, then there is an acceptable probabilistic approach. According to this approach, what we should concern ourselves with is solving one problem at a time: given some relatively small algebra, we define a probability function over it — perhaps in some systematic way, such as that suggested by Jaynes.[4] The probabilities are dependent on the way in which the problem is structured, but that is all right. We are looking for solutions to problems we happen to have. That someone else might structure the problem differently, and get different results from the same evidence, is quite irrelevant.

One would prefer it, of course, if there were compelling objective standards for approaching a given problem. None have become universally accepted so far. One response is to go ahead with what we have, subjective though it may be. Another is to seek, as far as possible, objective answers to these questions, even if these answers are arbitrary. Here we shall pursue the course of seeking a procedure that will render science and scientific inference objective. Part of the motivation is that unless we try to find these hypothetical objective, logical standards, analogous to those for deductive logic and mathematics, we surely will not find them. Another part of the motivation is that it is worth trying to find out how far objectivity can take us.

Acceptance

It goes along with the idea that the conclusion of an inductive argument is a categorical statement, and not a statement of probability, that inductive arguments may warrant our accepting their conclusions. This accords with common sense: when the evidence is overwhelming that virus V causes disease D, we feel comfortable telling others that that is the case; we include that statement in our medical textbooks; we have no hesitation in acting on that hypothesis.

But this conception is highly controversial in the philosophy of science, as we have just noted. Philosophers as diverse as Carnap[5]

---

[4]E. T. Jaynes, "On the Rationale of Maximum Entropy Methods," *Proceedings of the IEEE* **70**, 1982, 939–952.

[5]Carnap, loc. cit.

(who thought of induction as central to science) and Popper[6] (who doesn't believe in induction at all) agree that responsible scientists never accept the propositions we think of as comprising the body of scientific knowledge. According to Popper, these statements should serve only to focus our efforts at refutation; according to Carnap, we must regard these statements as (at best) probable relative to the evidence we have concerning them.

On the other side are arrayed an equally diverse group of philosophers, such as Levi[7] and Salmon.[8] Much of scientific knowledge, from this point of view, must be *accepted* in order to serve as evidence for the acquisition of further scientific knowledge. In order to learn the melting point of a new organic compound, we must take the laws and theories underlying the measurement of temperature for granted.

There are further difficulties, in addition to those mentioned in the last section, for either side of the debate over acceptance. Suppose we follow Carnap and regard our body of scientific knowledge as consisting of statements of the form: the degree of confirmation of the hypothesis H, relative to the evidence E, is p, or $c(H,E)=p$. On Carnap's view, these statements are *not* subjective, but logically true. They can nonetheless serve as a basis for action, since if E is our actual body of evidence, then our *rational* degree of belief, the degree of belief we *ought* to have in H, is p, and statements of this form suffice, combined with a utility function, to generate decisions.

But what about E? The sort of evidence that is cited in support of scientific laws and theories is not incorrigible; better experiments in the future may lead us to revise the evidence. So the "evidence" we cite—unless we cite merely the way things appear to us—must itself be regarded, according to one point of view, as "merely probable." If we look at the matter this way, we are led to a view like that of Richard Jeffrey[9]: our body of scientific knowledge is to be regarded as an infinite field of propositions, together with a probability distribution over that field. We can be led by experience to modify the probabilities of certain observational propositions; that modification then propa-

---

[6] K. R. Popper, "On Rules of Detachment and So-Called Inductive Logic," in Lakatos, op. cit., pp. 130–139.

[7] Isaac Levi, *The Enterprise of Knowledge*, MIT Press, Cambridge, Mass., 1980.

[8] W. Salmon, "Who Needs Inductive Acceptance Rules?" in Lakatos, op. cit., pp. 139–144.

[9] R. C. Jeffrey, *The Logic of Decision*, 2nd ed., University of Chicago Press, Chicago, 1984. (First edition, 1965.)

gates itself through the field in accord with the axioms of probability to yield a new probability distribution.

There are two serious problems with this view. First, there is the problem of the source of the probability distribution. Jeffrey is a subjectivist, so for him, the source of the distribution is just the beliefs of the agent. Of course, many agents have similar degrees of belief, so subjectivism may not ordinarily lead to very divergent opinions; nevertheless, when opinions *do* diverge, there is no way of resolving the conflict (except by gathering more evidence). But given any amount of evidence, any amount of disagreement is still possible. It seems that there is, or should be, more objectivity than this in science.

Second, the system is unworkably complicated. Not only must we find some way of specifying the probability function over the whole field of propositions with which we are concerned, but that probability function must be perpetually updated as new experiences impinge on the agent in question. It is far simpler to suppose that there is a stock of statements that are (generally) not in question in an inquiry or a decision-making context. Then there are other statements, not part of this accepted stock, whose probabilities we may consider.

Acceptance is not without its problems, too. Suppose that we accept any statement whose probability is greater than 0.9. Why 0.9? The *level* of acceptance seems arbitrary, and what is a plausible level of acceptance in one context may not be plausible in another context.

Suppose we accept a statement S into our rational corpus K if and only if its probability exceeds p. This leads to the following difficulty, known as the *lottery paradox*.[10] Consider a lottery that we know to be fair, and that has enough tickets so that the probability that a given ticket loses exceeds p. Let S_i be the statement that ticket i loses. For every i (before we know the winner), S_i has a probability for us that exceeds p and is therefore acceptable as part of our body of knowledge. We assumed that we knew the lottery was fair, and this yields the acceptability of the sentence "For some i, not-S_i." The set of acceptable sentences is not consistent.

Is this, as it has been taken to be, paradoxical? Should the set of acceptable sentences be consistent? It seems natural for me to believe, with reason, that one of the many statements I believe with reason is false. (Indeed, if I believe that, it is true automatically! But that joke

---

[10]The lottery paradox was first discussed in H. Kyburg, *Probability and the Logic of Rational Belief*, Wesleyan University Press, Middletown, Conn., 1961, p. 197.

depends on playing fast and loose with the object language/metalanguage distinction we have been taking for granted.)

More plausibly and persuasively, consider a large number of measurements of different objects performed by a process with a known distribution of error. It is certainly reasonable to believe, of each measurement, that the observed value is within three standard deviations of the true value. (If that is not secure enough for you, make it six standard deviations.) If the number of measurements is large, it is also reasonable to believe that at least one of the measurements is *not* within three standard deviations of its true value. Put otherwise, the long conjunction of the set of statements asserting of each measurement that it is within our standard of accuracy and the statement denying that conjunction is inconsistent.

The ticket in this measurement lottery will never be drawn. We have, we suppose, no reason whatever to single out any particular measurement to be maximally suspect, or indeed, any more suspect than any other. To suspend judgment, to say that we don't know the lengths of any of these objects we have measured, is skeptical defeatism at its worst, unless we are prepared to give up acceptance altogether and to retreat to the alternative view of scientific knowledge.

Suppose that we consider a body of knowledge K. Suppose that if statement S is probable enough (whatever that may mean) relative to K, we add it to K to get K'. This is surely the simplest approach to probabilistic acceptance. But one of the most essential features of inductive logic is its nonmonotonicity: the fact that what is probable or acceptable relative to one body of evidence may cease to be probable or acceptable relative to an *expansion* of that body of evidence. This approach would fail to be nonmonotonic; as we noted in the last chapter, if S is in K', its probability will generally be $[1.0, 1.0]$ relative to K' and relative to any expansion of K'. There is no getting rid of statements that have once been accepted.

Faced with these difficulties, we had better take a closer look at the notion of a rational corpus if we are to pursue the representation of scientific knowledge in an acceptance framework.

Bodies of Knowledge

The most important property of inductive inference is its nonmonotonicity, or self-correcting character. There is no point in regarding a proposition as uncertain if there is no way to correct it. Corrigibility is crucial. The most straightforward way to handle this is to consider two bodies of knowledge simultaneously: one is the evidential corpus — the set of propositions acceptable as evidence in a certain context; the other is the practical corpus — the set of propositions counting as "practically certain" in that context. Clearly the former will be a subset of the latter.

If we construe the practical corpus as consisting of exactly those statements whose probability relative to the evidential corpus exceeds some given level (about which more in a moment), then we have satisfied the demands of nonmonotonicity: statements will appear and disappear in this corpus as their probabilities, relative to the evidential corpus, wax and wane, that is, as the set of statements accepted as evidence changes. When we get more evidence, an acceptably probable statement may become unacceptable, and in fact its denial may become acceptable.

As observed in the previous chapter, one of the properties of probability is that if S entails T, the lower bound of the probability of T will be at least as great as the lower bound of the probability of S. Thus the set of practical certainties is weakly deductively closed: it contains the deductive consequences of every statement it contains. It is subject to the lottery "paradox" insofar as it may contain each of a set of statements that are jointly inconsistent. But it does not uselessly contain all statements, because it contains no explicitly contradictory statement. Nor does it contain both a statement and its denial, so long as the level of acceptance is chosen to be greater than .5, in virtue of the fact that $Prob(S,K)=[p,q]$ if and only if $Prob(\sim S,K)=[1-q,1-p]$; p and $1-q$ can't both be greater than .5. (In general, if the level of acceptance is set at r, it takes $n>1/1-r$ statements to imply a contradiction.[11])

The level of practical certainty is indeed arbitrary, though no more arbitrary than the corresponding values of $\alpha=.10$, .05, and .01 so popular in applied statistics. There are some intuitive considerations bearing on this level. Practical probabilities — the probabilities and distributions that enter into the computations of mathematical expectation for the purpose of evaluating decisions — are, naturally, taken relative to the corpus of practical certainties. If the level of practical certainty is r, practical probabilities greater than r or less than $1-r$ seem suspect in the computation of expected utility. Conversely, if we

[11]For further discussion of these questions, see H. Kyburg, "Conjunctivitis," in M. Swain (ed.), *Induction, Acceptance, and Rational Belief*, Reidel, Dordrecht, 1970, pp. 55–82.

need to take account of probabilities greater than r or less than $1-r$ in practical deliberation, we should set our level of practical certainty at a value greater than r.

(A practical example of the violation of this intuitive principle occurred when it was solemnly declared by some authority that the probability of a certain sort of nuclear plant disaster in California was less than 10^{-72}. The probability of the data on which this number is computed is surely less than $1-10^{-72}$.)

If the corpus of practical certainties is not deductively closed, how do we account for the vital role of deduction in scientific argument? Consider a deductive argument from premises $P_1, \ldots, P_n$ to the conclusion C. In order to have reason to accept the conclusion on the basis of this argument, we must have reason to accept the premises. But this maxim does not distinguish between having reason to accept *each* of the premises and having reason to accept their conjunction. If we think about uncertainty, it seems clear that the cogency of the deductive argument depends on our acceptance of the *conjunction* of the premises, however convenient it may be, in writing out the argument, to write the premises on separate lines. If we construe deductive cogency in this way, then it fits into our picture of the practical corpus: if the conjunction of the premises occurs in the corpus, then the conclusion should appear there in virtue of weak deductive closure.

What can we say about the evidential corpus? We speak of an evidential corpus because we wish to take account of the uncertainty that may infect even our statements of evidence, relative to which we judge the acceptability of statements for our corpus of practical certainties. This may seem baroque, but if we want to reflect scientific practice, we must allow for the ultimate rejection of even evidential statements. We can accomplish this by a shift of context. If the acceptance of a statement in the evidential corpus becomes an issue, we shift gears and regard that corpus as the practical corpus, and another corpus, of higher level, as the evidential corpus. (For playing skeptical philosophical games, we could even take the evidential corpus to consist of logical truths and phenomenological observation statements characterized by certainty—if there are any such.)

We support the uncertain by reference to the less uncertain. Thus it seems inevitable that the evidential corpus should be characterized by a higher level of acceptance than the practical corpus. Although we never have deductive closure in either our practical corpus or our evidential corpus, the following theorem follows from our conventions about probability: if S and T are in the evidential corpus, then if S is in the practical corpus, the conjunction of S and T will be in there too. This is a consequence of the sentential tautology $T \rightarrow (S \equiv T\&S)$. In view of the fact that in the classical probability calculus, if S and T are independent, the probability of their conjunction will be the product of their probabilities, this suggests that we take the level of the practical corpus as the square of the level of the evidential corpus, even though the result does not depend on "independence."

But what levels? The appropriate levels of practical and evidential certainty depend on context. We could let levels of acceptance vary continuously or have any real number value. There is little to be gained from this, however, and something to be gained from the ability to speak of the "next higher" level of acceptance. Furthermore, the representation is imperfect anyway; there is little to be gained by attempting to carry it out to too many decimal places. Let us take the levels to be numbers of the form $(1/2)$ raised to the power $(1/2)^k$ for integers $k \geq 1$. Put otherwise, if r is a level of practical certainty, then for some integer k, $\log_2 r = -k$, and the corresponding level of evidential certainty is $(r)^{1/2}$. Note that $\log_2 (r)^{1/2} = -(k+1)$. The sequence of numbers looks like this: .71, .84, .92, .96, .98, .99, .9946, .9973, .9986, .9993, .9997, and so on.

We have not discussed the question of observation statements. We shall consider them in more detail later but, loosely speaking, we will be able to characterize observation statements in terms of their reliability and add to a corpus, on the basis of observation, statements whose reliability exceeds the level of that corpus. (Note that this procedure cannot be construed directly in terms of probability as we have characterized it so far; we can't say that we know that a high proportion of observation statements are true, and therefore that this random observation of that type is probably true. But indirectly, we can provide an explication of reliability that comes to the same thing.)

Varieties of Induction

As a preliminary trial of the machinery just described, let us return to the various forms of inductive argument mentioned earlier.

Instantial induction, which is not generally mentioned as a form of induction at all, though it is clearly a species of nondemonstrative or uncertain inference, is straightforward. Suppose that the level of practical certainty is .98, corresponding to $k = -5$. Suppose that in the evidential corpus we have the statement "%$(B,A) \in [.99, 1.0]$" and the statement "$a \in A$." It is claimed that "$a \in A \& a \in B$" is among our practical certainties. This will be the case if and only if the probability of this conjunction, relative to the evidential corpus, is greater than .98. Consider the inference structure

"$a \in A \& a \in B$" $\quad a \quad A \quad A \cap B \quad [.99, 1.0]$

The question is, is a a *random* member of A with respect to $A \cap B$, relative to the evidential corpus? It may be, but it may also not be. It depends on what else we know. It is consistent with the description of the evidential corpus that it should contain the statement "$\sim a \in B$"; if so, then a is clearly not a random member of A in the sense required. But it is also consistent with the description of the evidential corpus that it should be. We cannot consistently say that "all we know of a is that it belongs to A," for we know that it belongs also to $A \cup C$ for any C; we know that it belongs to its unit set; and so on. But if everything we know about a is a logical consequence of "$a \in A$," or of other statements in K that don't mention a, then a will be a random member of A in the appropriate sense, the probability of "$a \in A \& a \in B$" will be $[.99, 1.0]$, and "$a \in A \& a \in B$" will be a member of the corpus of practical certainties. An instantial inductive argument will have shown that it is a member of this corpus.

Statistical induction is strictly analogous to instantial induction. As we saw in the last chapter, it is easy enough to have statements in K of the general form: the proportion of samples that are representative (in a certain technical sense) of the parent population is high. These statements may appear in inference structures, and *if* a specific sample is a random member of the set of samples with respect to being representative, relative to the evidential corpus, and the proportion is high enough, we may *accept* in the corpus of practical certainties the sentence "Sample a is representative." But sample a is representative if and only if the parameters characterizing the parent population satisfy certain numerical constraints expressed in terms of the sample parameters. So we may also accept in the practical corpus the statement that the parameters of the parent population satisfy these constraints.

The problem of generating the appropriate inference structures is a problem in classical statistics and introduces no special difficulties. Some such inference structures represent strictly mathematical (set-theoretical) facts, such as the fact that almost all subsets of a set reflect the frequency with which members of that set belong to some other set. Some represent empirical facts as well, such as the fact that though we know nothing about the mean and variance of the quantity Q, we do know that Q has an approximately normal distribution. Some may involve yet more complicated empirical facts, such as the fact that we know that Q has a different (perhaps approximately) known distribution in each of the sets $A_1, \ldots, A_n$, that A_i is selected with a certain (perhaps approximately) known frequency, and that a sample has been selected from one of the sets $A_1, \ldots, A_n$. (This is an abbreviated way of describing the classical Bayesian inference.)

As is generally the case, the complicated logical and epistemological issues concern not the generation of inference structures for statistical statements—though this is surely difficult enough—but adjudication among these competing inference structures in terms of randomness or (equivalently) the determination of the appropriate *reference class*.

To illustrate some of these issues in a classically and conventionally abstract way, suppose that we are confronted with a set of urns, each containing black and white balls. We choose an urn and choose a sample of n balls from it (with or without replacement; the arithmetic is different, but the principles are the same). As outlined in Chapter 3, for suitable r, we can show that the frequency with which n-membered subsets of the set of balls in the urn are $R(r)$—that is, representative in the sense that the relative frequency of black balls in the sample differs by less than r from the relative frequency of black balls in the urn from which the sample was drawn—is greater than *prac*, the level we have adopted as practical certainty. Thus we may accept among our practical certainties (characterized by an acceptance level of *prac*) the statement that the particular sample we have taken is $R(r)$ or, equivalently, that the frequency of black balls in the urn lies between $m/n-r$ and $m/n+r$.

Now observe that exactly the same argument goes through if we consider our sample of n balls as a sample of balls in the urns in general. The fact that all these balls came from the same urn does not undermine the inference from the composition of the sample to the conclusion that the frequency of black balls among all the balls in all the urns lies between $m/n-r$ and $m/n+r$. No more does the fact, in the first argument, that all of the balls comprising our sample were *drawn* from the urn (as opposed to *being* in the urn), or the fact that all were drawn by a female American, undermine that inference.

Our instincts cry out against this conclusion. But that is because we know that in such examples (*are* there any urns in the real world?) the urns contain different proportions of black balls. An enlargement of our evidential corpus to reflect this kind of knowledge would change things. Suppose that we know that $k_1\%$ of the urns had $j_1\%$ black balls, $k_2\%$ of the urns $j_2\%$ black balls, and so on. Or, more generally, we might know that a fraction between w_i and v_i of the urns have a fraction of black balls between p_i and q_i. This gives us a new inference structure that concerns *pairs* consisting of an urn and a sample of n from that urn, and that contains a statistical statement corresponding to the Bayesian analysis. This new inference structure may or may not *differ* from our old one, according to whether our prior knowledge (of the w_i, v_i, p_i, q_i) is more or less precise. If the new inference structure does differ from the old one, in our technical sense, then it will dominate the original inference structure and be the appropriate inference structure for inductively inferring the proportion of black balls in the urn we have chosen. Note that it will no longer give us a basis for inferring the *overall* relative frequency of black balls.

Universal induction has been thought to be the form of inference typical of sound science: we observe some A's, all of them turn out to be B's, and we infer that all A's are B's. On the view being propounded here, this is not so. We can, under appropriate conditions, infer with practical certainty from the fact that all the A's we have seen have been B's that *nearly* all A's are B's. But this is just statistical induction. How do we go the rest of the way?

There are two cases to be considered. The first case has more of an empirical flavor; the second involves what I have called *nomic* induction. For the first case, suppose that our evidential corpus contains the statement that of a sample of n A's, n have been B's. This warrants (if the appropriate conditions of randomness are met) the inclusion in our corpus of practical certainties of the statement "$\%(B,A)\in[1-\delta,1.0]$," where δ is some small positive number. Suppose that δ is smaller than $1-$the level of the practical corpus.

We add one more constraint: that the sentence "$(\exists x)(x\in A \,\&\, \sim x\in B)$" does *not* belong to the practical corpus. Now shift gears and regard the corpus of this level as the evidential corpus. For any term a, if "$a\in A$" belongs to this (new) evidential corpus, "$a\in B$" will belong to the corresponding (new) practical corpus. This will be so because either, relative to the evidential corpus, a will be a random member of A with respect to B, and thus the probability of "$a\in B$" will be $[1-\delta,1.0]$, which is high enough to warrant the inclusion of "$a\in B$" in the practical corpus by instantial induction, or a will be a random member of B (if we know that $a \in B$ in the evidential corpus) so that the probability of "$a \in B$" relative to the evidential corpus is [1.0,1.0], which again is high enough to warrant the inclusion of "$a \in B$" in the practical corpus.

Exactly the same results are obtained by including the universal generalization "$(\forall x)(x \in A \rightarrow x \in B)$" in the evidential corpus. It allows us to infer "$x \in B$" from "$x \in A$." This fact may underlie the intuitions of those philosophers, like Ryle[12] and Toulmin,[13] who have argued that universal generalizations should be construed as rules for making inferences, rather than as sentences in the object language of science. As Cooley[14] has persuasively argued, this revolution in itself does not buy much. But it does reflect one way in which universal generalizations may function instrumentally at certain stages of scientific inquiry.

Nomic generalizations are a different kettle of fish. The traditional difference between merely universal and nomic generalizations is that nomic generalizations support counterfactuals, while mere empirical universal generalizations do not. If the denial of the nomic generalization is analyzed as "it is possible that there should be an A that is not a B," and we analyze that in turn as "'$(\exists x)(x \in A \& \sim x \in B)$' is consistent with the evidential corpus," then it is true that the universal generalization equivalents we have considered do not support counterfactuals. We have only stipulated that "$(\exists x)(x \in A \& \sim x \in B)$" not, in fact, appear in the evidential corpus, and not that it be inconsistent with it. Note that our universal generalization equivalents *do* support hypotheticals: if we add "$a \in A$" to the evidential corpus, we may infer "$a \in B$" in the practical corpus.

I shall construe nomic statements generally as characteristic of a particular language, and thus as "generalizations" in virtue of their universality, rather than in virtue of their source in experience. The same will be true of theories. Thus statistical and universal induction and nomic and theoretical induction will be dealt with along very different lines: I shall take both laws and theories as conventions of our scientific language. But I shall take conventions not as arbitrary, but rather as subject to rational assessment in epistemic terms. Thus

---

[12]G. Ryle, "Predicting and Inferring," in S. Korner (ed.), *The Colston Papers*, Vol. 9, Butterworth's Scientific Publication, London, 1957, pp. 165–170.

[13]S. Toulmin, *Foresight and Understanding*, Indiana University Press, Bloomington, 1961.

[14]J. C. Cooley, "Toulmin's Revolution in Logic," *Journal of Philosophy* 56, 1959, 297–319.

the proposal to take our scientific language to be characterized by one conventional law rather than another (or by a conventional law connecting certain quantities, rather than by no law) is a proposal to be evaluated in light of the evidence. The result of this evaluation will not be a measure of evidential support or probability enjoyed by the proposed law, but it will nevertheless involve probability indirectly. We will not be able to say that, relative to the evidence we now have, Einstein's theory of special relativity is probable, but we will be able to say that probabilistic considerations support the preferability of a language embodying Einstein's theory over a language embodying Newton's theory.

To develop this proposal requires that we consider some other matters first. Basically, however, we will have two quite distinct forms of induction: on the one hand, there are instantial and statistical induction and the use of universal generalizations in a nonnomic sense; and on the other hand, there is the choice, informed by empirical data and reflecting the total state of our bodies of knowledge, between languages embodying various nomic and theoretical conventions.

5

Observation and Error

Experience and Judgment

A central empiricist claim is that it is from experience, and only from experience, that we obtain empirical knowledge about the world. Both "experience" and "knowledge" deserve discussion.

Hume distinguished between knowledge of matters of fact and knowledge about the relations of ideas. It is the former, rather than the latter, that is alleged to derive from experience. The truths of mathematics, for example, or of logic itself, are construed as depending on the relations of ideas. In the framework we have been developing, "ideas" are represented by the terms of our formal language. Their relations, then, are embodied in the axioms of the language: That "Pa" implies "$Pa \lor Qa$," that "$Pa \lor {\sim} Pa$" is a theorem, that "$2+3=5$" are all consequences of the logical and mathematical axioms of any typical formal language. A language formally representing a fragment of ordinary discourse may also contain axioms from which it follows that all bachelors are unmarried.

It will be argued in due course that many of the statements that we might think of as formal analogs to statements of matters of fact are best construed as embodying features of the (formalized) language of science. This argument need not concern us now; what I wish to point out is merely the existence, in our bodies of knowledge, of sentences that depend on the axioms of our formal language and of sentences that do not.

One way of looking at the special sentences that follow from the axioms of our language is to construe them as "analytic," or, more linguistically, as sentences that embody prelinguistic analytic truths. This notion of analyticity is due essentially to Kant. Since a bachelor is an *unmarried male*, "All bachelors are unmarried" is analytic. Its truth follows from an analysis of the concepts involved. But analyticity is construed more broadly nowadays; for example, "The length of the collinear juxtaposition of two rigid bodies is the sum of the lengths of the rigid bodies" might be considered analytic, even though it is hard to see how to fit it into the pattern of "All bachelors are unmarried."

For present purposes, however, we would do well to ignore analyticity, in either a broad or a narrow sense, and to focus on theoremhood in a formalized language. That is because our concern is to provide a formal model reflecting human knowledge and belief in a rational scientific context. We adopt in our formal language axioms that imply the truth of "All bachelors are unmarried," because any reasonable representation of human knowledge would embody a statement corresponding to this statement.

Knowledge of the theorems of a formal logic is a species of knowledge; we may call it logical knowledge. Experience may nevertheless be required in order for a person to *achieve* logical knowledge; he may need to construct or be shown a *proof* of the theorem. This is experience of a different sort than that on which we base our knowledge (for example) that about 51% of live human births are the births of males. It is a variety of experience that appears to depend on the formalizability (if not the formalization) of certain (logical) aspects of our language. It also appears to be the case that only organisms with certain predispositions and propensities are capable of achieving this kind of knowledge, and it may well be the case that these predispositions and propensities impose certain constraints on the structure of any learnable language. This again is not a matter that is directly germane to our concerns with formalizable scientific knowledge. Linguistic and logical knowledge, and the sort of experience that leads to it, are not the sort of knowledge and experience the empiricist has in mind when he alleges that all knowledge is founded on experience.

We are concerned with empirical scientific knowledge. Let us see if we can characterize this a bit more precisely. Given a formal, axiomatized, scientific language, we can specify formally the set of sentences that are provable from those axioms. (This is smaller than the set of *valid* sentences that we can specify semantically, but the difference need not detain us.) These sentences should be included among the sentences that we are *committed* to; this is not to say that we should actively believe them, or even that, to be rational, we should assent to them when asked, but only that a proof from axioms should lead us to acknowledge our commitment. Such sentences will be said to appear in our (ideal) rational corpus of knowledge.

But these are clearly not the only sentences we are justified in believing. We are justified in believing particular sentences ("*a* is a crow"); statistical generalizations ("About half of all coin tosses yield heads") and possibly universal generalizations that are not entailed by the axioms of the language, "All crows are black" being a frequently alleged example; and perhaps even general physical or biological theories. The empiricist claim concerning knowledge, then, is the claim that every sentence that is not entailed by the axioms of the language and is justified is justified directly or indirectly by experience.

"Experience" is a notoriously vague term. What the empiricist has in mind, though, is not so obscure. He means "sense experience" — that is, auditory, visual, tactile, olfactory, kinesthetic experience. Here empiricists fragment. One tendency has been to seek ever more "secure" foundations for knowledge. If one might be mistaken about whether such and such a bird was a crow, we had better take our directly justified belief to be the belief that such and such is a bird with such and such visual characteristics. If one might be mistaken about whether or not it is a bird, perhaps we would simply affirm the existence of a certain sort of sense datum. But entirely aside from the difficulties of founding our scientific knowledge of the world on sense data — difficulties that will be displayed in due course — there is the fact that we do not ordinarily observe, or experience, or even affirm the existence of, sense data. It is questions of psychological or metaphysical theory that lead us — if anything does — to sense data. In the quest for certainty and security, we have historically been led to take our observational experience to be more and more private, and less and less like the public evidence, available in principle to all, on which we like to think our scientific knowledge rests.

But if we resist the pull of security in the foundations of our knowledge, we must always acknowledge the possibility of error in our observations. And how can we make sense of that, since there is no way of matching our experience directly against reality? How can we tell whether or not *all* of our observations are erroneous? Or some particular class of them? And then there is the problem of how far to go in taking observations at face value. Is my observation of a book, mediated by contact lenses, "direct"? Or spectacles? Or a telescope or a microscope? Or a television set? When I see a white streak in the sky, am I seeing an airplane? Or only a contrail? Or only a white streak in the sky? Or a white datum against a blue datum?

What we need, for our purpose of understanding the nature and content of scientific knowledge, is not so much a solution to these problems as a way of avoiding them. The key is the somewhat old-fashioned-sounding notion of *judgment*. More specifically, I shall suppose that in our formal language we can specify a certain set of sentences that are the *kind* that can represent the content of an observational judgment. Why just *those* sentences? We can say that it is just those sentences that are picked out as observational by the language itself; that is, part of specifying a language is specifying (in some suitably recursive manner) the set of *observation sentences*.

This means, of course, that we may have two languages, otherwise identical, that are distinguished by the fact that the sets of observation sentences picked out in each are different. We may nevertheless (as we shall see later on) have *good reason* to adopt one language rather than the other. In the present context, that means that we can have good reason to take one set of sentences as observational rather than another.

We will also discover grounds for assessing the *reliability* of observation sentences of various sorts. This is a matter closely related to the grounds for choosing between languages: if the observation sentences of one language are so unreliable that judgments involving them are wrong more often than not, and the other language is not that much of a disaster, then the other language is clearly preferable.

We can still say something general about the characteristics of observational sentences. For one thing, they are (or are intended to be) the sorts of sentences that we can *sometimes* judge to be true or false on the basis of what happens to us. Such judgments need not always be veridical; we must allow the possibility of errors of observational judgment in the best of worlds and with the best of languages. For another, they are (or are intended to be) the sorts of sentences that admit of publicly uniform judgment. This publicity has to be hedged around with considerations of perspective, training, attention, and the like. Nevertheless, the public function of language as a medium of communication requires some degree of publicity among some of the sentences of the language, and this must already be paid for in the possibility of error.

Once we acknowledge the possibility of error or untruthfulness, the way is open to us to consider that class of observation statements on which most of our knowledge of the world is based: the written or oral testimony of others. Observing someone to say something is not what some people have in mind when they advocate "learning by experience," but it surely must be taken to satisfy the empiricist's demand that knowledge be based on experience. Of course we must take account of the possibilities of lies and errors and misunderstandings, just as, in the case of other kinds of observation, we must take account of misperception, hallucination, poor vision, lack of light, and the like. And of course different circumstances will be characterized by different error rates.

What is new and interesting about "testimonial observation" is that it is so closely tied to what we might call language instruction. "There is a crow on the barn" could serve either function: if you can see the bird as well as I can, the purpose of my utterance is probably to augment your ornithological vocabulary. If you are blind, or looking the other way, the purpose of my utterance is (probably) to augment your corpus of empirical knowledge. In the first case you take my utterance to concern the relation between the word "crow" and a particular feathered object (to the effect that the latter is an instance of the class of objects to which the former is applicable); in the second case, you take my utterance as testimony concerning the state of the world.

Sometimes it can be quite difficult to tell what is going on. Suppose you point out a blue bird to me and say, "There's a crow," calling my attention to it. I answer, "No, you're wrong. All crows are black, and that bird is blue." Am I correcting your use of the word "crow" on the grounds that it is incorrect to apply it to nonblack objects, or am I informing you of a (generic) fact about the world? Or am I correcting what I take to be your perceptual claim that the object in question is black? If I am informing you of a generic fact about the world, how is it that I have the hubris to rank my generalization above your counterexample?

We see already that the relations among judgment, error, and generalization are going to be far from simple. The Popperian formula, "Put a generalization to the test by looking for a counterexample," is plausible only when observations are taken to be incorrigible. If we take observational judgments to be corrigible, it may take more than a single observation to refute a generalization. (How many? It depends.) Testing is still possible, of course, even if there is no definitive test. And in some cases (with certain background knowledge), we can approach a definitive test. And in a sense, testing becomes even more possible if we admit the uncertainty of observation, since if we admit corrigible observation judgments as "observations," the scope of observation is greatly increased.

The Scope of Observation

Ordinary language provides the first cut at picking out the observation sentences of our formal language. "Longer than," "round," "red," "heavy," "sweet," "stinky," "alive," "between," "angry," and their ilk are predicates that may perfectly well be represented in our formal language, and they are predicates that may be predicated of appropriate objects on the basis of observational judgment. (We may contrast not only "electron" and "magnetic field" but also "bachelor" and "unmarried.") Such judgments may sometimes be in error—an elliptical penny will look round to the ordinary observer[1]—but in general (as we shall see), they are reliable. Furthermore, such judgments admit of a high degree of public or interpersonal uniformity: in similar situations, most people will come to form roughly the same judgments. Most people can tell when someone is angry, even if he is not from their own culture. In fact, such judgments can be reliably made across species. This uniformity is perhaps "instinctive"; it is inculcated in the learning of a language, and it can be further enhanced, in special cases, by explicit training. It is also, of course, precluded in special cases: a person may be blind or deaf.

The results of practice and training should not be confused with inference. To know of a certain animal that it is angry, for example, may require some familiarity and experience with animals of that species, or closely related species. People with that familiarity can tell dependably when that kind of animal is angry. (The same is true of other species; a dog can easily learn to tell when a cow is angry.) It might be maintained that this is a matter of *inference*, just as it is sometimes (less plausibly) maintained that judging that something is about 100 yards away, or that it is square, is a matter of inference.

If it is to be called "inference," however, it must be construed as unconscious inference. It is no doubt true that there is a psychological process that might be called "inferring," leading to these judgments

---

[1] It takes a *trained* observer to see a round penny viewed from an angle as an ellipse.

and there is no reason that this process cannot be unconscious. Indeed, there may be good reasons, in exploiting the analogy with "inference" as it is construed in logic, for taking the process to be inferential in character. One might also have good reason to speak of "inference" in characterizing the process by which "information" is transformed and transmitted on its way from the rods and cones of the retina to the visual cortex, or in characterizing the process by which a computer transforms and transmits information gleaned from its television eye to some other program as "inference." This can be a sensible way to talk when we are postulating an explicitly inferential mechanism to account for the transfer, as we might in the case of the computer program. But under ordinary circumstances, I *see* a cow and *infer* that it has escaped from somebody's pasture; I do not *see* a large black spot on my visual field and *infer* that I am confronting a black angus cow. In fact, I see a cow and *infer* that there is a distinctive area of my retina that is responding to light reflected from the cow.

We should distinguish here between the psychological (or physiological, or computational) process, and the logical relation between linguistic entities. The latter may be reflected in our formal language and is subject to critical control. The former is (more obviously) built into the organism or the computer. Where to draw the line may not be clear, but at some point, one must accept sentences of one's formal language on the basis of direct judgment.

This is not at all to deny that there are behavioral clues (and perhaps clues in the form of imperceptible odors and sounds) to the state of mind of an animal. Nor is it to deny that we can, through inquiry or sometimes through reflection, discover what those clues are, though to do so, particularly in such a case as the one at hand (the angry cow), may be far from trivial. (Note that we can relatively easily find some correlation, such as that a cat, when angry, often lashes its tail back and forth; but to find such a correlation is just to find a correlation between the epistemically *independent* judgments of tail lashing and anger.)

It is also possible to replace a language in which the term "is angry" is observational by a language in which it is not. There are a number of motivations for doing this. The soundest motivation would be that "anger" has come to play an important role in a web of theoretical terms only indirectly related, as a whole, to more conventional observational terms. The least sound motivation (since we can always get by without the replacement) is that we want to make use of untrained observers. Thus, if we are doing a study of anger in the kinkajou,[2] we had better give our graduate assistants a list of behavioral criteria that are appropriately related to the kinkajou's state of mind. They are unfamiliar with the animal, and though *we* can tell when a kinkajou is angry, their reports will be more reliable if they *infer* anger from direct observations of behavior. But then, should we not just have them report the behavior?

This example contrasts interestingly with a corresponding example from histology. It takes experience with a microscope, and with applying stains, to learn to identify certain sorts of cells. (Perhaps as much experience as it would take to learn to identify anger in the kinkajou!) But we don't offer our graduate students in histology an easy out (perhaps because it would be hard to do); instead, we make them stare through their microscopes at slide after slide until they become capable of making reliable judgments of the kind called for.

Other examples come easily to mind of the importance and relevance of practice in making observational judgments. It is said that Eskimos distinguish 17 varieties of snow. I assume that they can do so reliably. Of course, there are clues that they use; some may have to do with the textures and qualities of the snow itself; others may have to do with the temperature and weather that produced the snow. No matter. So far as we need be concerned, the classification of snows is, for the Eskimos, a matter of observational judgment. A similar story can no doubt be told of the Polynesians' recognition of a large variety of wave forms. A more industrial example can be found in the machine shop. A tool and die maker can recognize various compositions of steel that look, to the beginner, indistinguishable. A romantic example: the Indian guide who sees the trail of the deer through the woods. It is the modern novelist who feels the need to mention the bent twigs, and overturned leaves; the Indian simply observes the path made by the deer.

If we take this attitude toward observation—namely, that there is a certain (quite large) set of sentences in the language that can justifiably be judged true or false on the basis of observational experience—then nothing stands in the way of taking observations through contact lenses, through microscopes, and through telescopes (even infrared sensing telescopes or radio telescopes) to be "direct" observation. The

---

[2] A nocturnal mammal native to Mexico, Central America, and South America, noted for its prehensile tail.

fact that a certain amount of machinery is used no more subverts the directness of the observation than does the use of a meter stick, or a vernier caliper, or a micrometer subvert the directness of the measurement of length.

Not everything about observation immediately becomes clear. In a cloud chamber it seems reasonable to say that we can observe the trajectory of a particle, but not, perhaps, the trajectory of an electron, even though what we *see* is a collection of droplets. But is it so very different for histologists to say that they see a certain kind of cell? How about a magnified photograph of the cloud chamber? We see (anyone can see) white streaks on the photograph, but is it incorrect to say that the trained particle physicist sees here the track of an electron, there the path of a positron, and so on?

The basic criterion I am suggesting has to do with the interpersonal uniformity with which people trained in the discipline in question can affirm or deny sentences on the basis of observational judgment. It is also related to the question of whether there is some *more* reliable way to get at those sentences. If the judgments are relatively reliable and relatively consistent from trained observer to trained observer, there is little more we can ask. This leaves three questions to be considered.

First, there is still some ambiguity about *inference*. It may be that when I see a cloud of dust, I judge that I am observing a herd of hartebeests, and that my African peers do the same. But the reason is nothing intrinsic to the cloud we observe; it is rather that in the part of Africa in which we live, it is only hartebeests that raise such a cloud. Might we not prefer to say that we observe a cloud of dust and infer that it is raised by a herd of hartebeests?

The answer is that we would. But our preference is based on our knowledge that very similar clouds of dust are raised (say) by herds of buffalo on the western plains (or herds of ATV's). Transport the African watchers to the plains of South Dakota, and we would expect him to be confused and his judgment to be unreliable. And since it is quite open to us to decide what sentences are to be regarded as subject to observational judgment, and what we should regard as inferential, it is quite open to us to use this knowledge as a basis for preferring the language in which we distinguish between the (direct) observation of the dust cloud and the (indirect) observation of the herd of hartebeests to the language in which the dust cloud–mediated observation of the herd of hartebeests is taken to be direct.

The second question concerns the observational judgments that trained individuals can make reliably, and with interpersonal uniformity, of objects that do not have the character that they are claimed to have. The practitioners of palmistry and phrenology can certainly agree on the lines they see and the bumps they feel, and they can, through practice, achieve far more agreement about these matters than the amateur could achieve. What distinguishes them from our histologists?

With regard to this question, there is no need to deny that palmists and phrenologists observe the lines and bumps that they claim to observe. Like the Eskimo who can see 17 varieties of snow, their observational prowess (we may charitably assume) has been developed along lines that ours has not. The question remains: what can they do with these observations? And we have every reason to believe that the answer is, very little. There is no well-substantiated and useful theory of either phrenology or palmistry. And so, the niceties of vocabulary employed to describe fine differences in heads and hands are of no use to us. We have no reason to enlarge our observational vocabulary to include those terms, and no reason to undergo a long apprenticeship to learn to use them correctly.

The third question concerns observational claims that we have every reason to believe are imaginary. One might have claimed, for example, that one could *see* the phlogiston escaping from the heated metal; certainly many have claimed to have seen ghosts and monsters. (Some have even claimed to have observed black swans!)

The answer to this question, appropriate to the observers of ghosts, is more interesting than the answer to the phrenologists. Seeing is a matter of attention and intention, as well as passive reception. This is a matter of well-justified scientific belief. It is also a matter of well-justified scientific belief that many of the things alleged to be true of ghosts cannot occur consistently with what we take to be our knowledge of the physical world. It follows that *if* we accept the physical "theory" we do (in quotation marks only because it is rather vague and general), *then* the ghostly observations must be regarded as erroneous, whatever the convictions of those who have made them.

What is interesting about this is that exactly the same sorts of considerations apply *within* physics. When a high school student gets a value for the acceleration of gravity that is at variance with the expected value, we have no hesitation about saying that her "observations" were in error. It would be *inconsistent* with accepted physical theory for her observational judgments to have been veridical.

There is no difference in principle, according to this way of looking at the matter, between recording the existence of luminous ectoplasm in a certain cemetery at a certain time and recording data from a pendulum experiment that yields a gravitational force at the surface of the earth of 45.7 feet per second per second.

If we allow that observation may admit of error without being thereby wholly undermined, then it is possible to take the scope of observation to be extremely broad. We need not seek to narrow the notion of observation so strictly as to render its results incorrigible. We can be tolerant of observational claims precisely because no observational claim is guaranteed to be free of error, and we can use our knowledge of error to distinguish observations of various degrees of reliability. But to make this approach work, we must be able to employ a quantitative, probabilistic, theory of error. And in order to remain true to our empiricist principles, this theory of error must be based on observational experience.

Theories of Error

One approach to the problem of error, already hinted at, is to construe the set of sentences that can represent observational judgments ever more narrowly in the hope of arriving at a characterization of observational judgment that allows observation to be error free. There is indeed a general way of doing this: we take observation to be of appearances rather than of realities. But then in addition, we must construe the appearances to be appearances to *us*. They become subjective and private. We lose our hold on the interpersonal and public aspects of observation.

Philosophically, this may be a feasible approach, though I rather doubt it. It is truer to the practice of science, as well as to the ordinary prejudices of common sense, to take our observation sentences to be objective statements about the world, and to admit that they are all subject to error.

Given that our observations are subject to error, how can we go about dealing with that error? It would be nice if we could simply delete the results of our erroneous judgments from our bodies of knowledge; but to be able to do that would be tantamount to being able to make error-free observations. That is not one of the options open to us.

We can tell more reliable from less reliable *kinds* of observational judgments. Poor light (or poor eyesight) undermines the reliability of visual judgments. Good illumination, spectacles, or a telescope, can enhance the reliability of visual judgments. These things seem like truisms, but what do such claims amount to? And can we justify those claims? We cannot, after all, compare our naked visual judgment with reality, and then compare our telescopic visual judgment with reality, and then argue that most of the time the latter corresponds better with reality than the former.

One extreme possibility would be to say that since we know that observational judgments may be erroneous, we should simply reject all observations. That would be silly. On the other hand, as we just noted, we often can't find a particular set of observations to pin all our errors on; otherwise, we would have the possibility of error-free observation.

Let us begin by asking how we know that some of our observations *are* in error. Clearly it is not by comparing our observation with the object observed. Nor is it merely by directly comparing one observation with another. If I observe at one time that a counter is blue, and at another that it is red, is it not possible that it has changed color? Isn't it possible that "it" is not the same counter? If I observe that an object is blue, and you observe that it is red, may this not be a reflection of our perspectives, or a misunderstanding about what "it" denotes? Even discounting temporal and perspectival matters, why should we not say both that the counter is red and that it is blue? The answer is that it is *impossible* for something to be both red and blue at the same time and in the same respect. Discounting temporal and perspectival matters, one judgment or the other is in error.

What is the source of this "impossibility" that reveals observational error? That can be a deep philosophical question if it is pursued far enough (or perhaps even a biological question), but fortunately, from our perspective, we can avoid that pursuit. Whatever its ultimate source, we can *reflect* the fact of this impossibility in the rules of our formal language. Suppressing mention of times and perspectives, we just take as an *axiom* or *meaning postulate* that "red" and "blue" are exclusive predicates when it comes to uniformly colored regions.

More generally, and leaving aside (for now) the question of the source and character of our general knowledge, as opposed to the observational contents of our rational corpus, it is the conflict between general knowledge and observation that reveals to us that sometimes our observations are in error. It is part of our body of knowledge that rigid bodies under constant conditions have a single determinate length. Suppose we measure a rigid body under constant conditions a number of times. We will (typically) obtain a number of different length readings (reports). The re-identification of the body, its rigidity, and the constancy of the conditions are all observational, as is the result of each measurement. Where do we locate the error?

That depends on the particular circumstances, but we will generally suppose that it is the measurements that are subject to error. Furthermore, under the right circumstances, we not only learn that the measurements are subject to error, but we obtain knowledge, by statistical inference, of an approximate *distribution* of that error. (Large errors occur less frequently than small errors; positive and negative errors occur approximately equally often over a normal range; and so on.) We shall look specifically at measurement and errors of measurement in the next chapter.

Let us first consider a qualitative example. Suppose that I have in my rational corpus (on whatever grounds) the knowledge that all crows are black. (I mean this literally: not that all normal crows are black, not that all but albino crows are black, but simply that *all* crows are black.) Suppose that I observe 100 birds that I identify as crows, and I identify the colors of 90 of them as black and 10 of them as nonblack. There is a conflict between the generalization and my observations. Error lurks *somewhere*.

Since I have supposed that I *know* that all crows are black, the error must lie in my observations. Either I have misidentified some birds as crows that really weren't crows, or I have misidentified some colors as nonblack. Of course, I may also have misidentified some crows as being other kinds of birds and some nonblack colors as black, but nothing in the example gives us any evidence concerning these errors. And I may have misidentified both the kind of bird and its color in some of the apparent counterexamples to the generalization. In any event, I know that at *least* 10 of the 100 observations incorporate some kind of error.

But how many errors are there, and where are they? They are in the pairs of observations that conflict with the known generalization, and they may also be in other places. *On the basis of the information we have*, there is nothing more we can say. I am assuming, of course, that we have no background knowledge of error rates, that these observations are all the observations we have, and so on. (*Of course*, this is unrealistic!) In a given case of an apparent crow that is apparently not black, we have no way of knowing whether it is the judgment that the bird is a crow, the judgment that it is nonblack, or both, that are in error. How do we go about purging the observational part of our rational corpus of error, then?

The answer is that we don't. But this doesn't matter. Surely no one maintains that they have never made a mistake, and no one should maintain that they are not making mistakes in their present state. So all of us have excellent grounds for thinking that there is some error in the set of items that we (dispositionally) believe; and, of course, if we could locate that error, we would expunge it. So all we can say is that there is some error (or errors) somewhere, and go on about our lives as if any *particular* belief were veridical.

Acknowledging the existence of error, even among the things we take most strongly to be true, need not paralyze our powers of deliberation, judgment, or argument. But it does make it important to treat error quantitatively and to control it.

Suppose that we know that judgments of nonblackness are about 90% reliable, that is, that about 10% of them are wrong. We judge that a given object is black. I take the content of the judgment to be categorical, that is, to be that the object *is* black, not that it is *probably* black. But there are still two epistemic alternatives. We may *accept* at a certain level of certainty (e.g., .90) an observation statement whose content is that the given object is nonblack. Or we may assign a probability of (about) .90 to that categorical observation statement. I have already argued in favor of the acceptance of inductive hypotheses. The same arguments apply here. And the same consequences follow. If we accept a large number of nonblack judgments, we can be practically certain that about 10% of these judgments are erroneous. *And we have no way of identifying which ones they are.* (Obviously, since if we have grounds for supposing a particular judgment to be particularly unreliable, we won't accept it, and our success rate will go up.)

Similar considerations apply to our "is-a-crow" judgments. If we suppose that the reliability of such judgments is also 90%, and that errors in making crow judgments are independent of errors in making nonblack judgments, then we will reasonably suppose that, among a lot of judgments, 10% will be in error concerning crowness, 10% will be in error concerning nonblackness, and 1% will be in error concerning *both*. Thus the number of errors out of 100 observations should be expected to be $.10 + .10 - .01 = .19$.

If we are looking for an *explanation* of our observations, this analysis seems appropriate.[3] But we have an explanation of our observations in these terms only if we *already* have the error rates from which to probabilistically deduce the likelihood of the observed error rates. That, however, is exactly what we lack in the primordial circumstance being envisaged—that is, when we lack background knowledge of error rates. In fact, it is exactly the effort to find grounds for such rates that we are concerned with here.

What basis do we have for getting at these error rates other than the rates of error demonstrated by our observations? And in the example at hand, the demonstrated overall frequency of error is exactly 10%. We are free to suppose that we have made more errors, but that conclusion is not forced upon us. It is also not forced upon us that the errors of each kind are independent of each other. So let us suppose that the basis from which we will estimate error rates is just the observed error rate and no more. There are still a number of things to consider.

Representing Our Knowledge of Error

It is hard to deny that we can sometimes be mistaken about our observations. Indeed, the pervasive possibility of error is one of the underlying arguments for skepticism. But to acknowledge the possibility of error, even in observation, need not entail abandoning the effort to give a formal epistemological treatment of observation. The formal framework we have already introduced, elevated to a metalinguistic level, will allow us to give a quantitative treatment of error.

Mary Hesse[4] introduced the important distinction between *observation statements* and *observation reports*. An observation report simply records an observation; it is, for example, the sentence written in a scientist's laboratory journal. "I observed a crow at about 8:17 A.M., 11/10/84, and it was black" and "Sample 3 weighed 3.456 grams on scale 2 on the first weighing, 3.449 grams on the second weighing, and

---

[3]This has been pointed out by Paul Scatena.
[4]Mary Hesse, *The Structure of Scientific Inference*, University of California Press, Berkeley, 1974.

3.460 grams on the third weighing" are examples. As the second example shows, if these statements were interpreted objectively, they would be inconsistent. The same object cannot weigh both 3.456 and 3.449 grams. These statements are interpreted rather as *records* of our judgments. This does *not* mean that they are "subjective" in any pernicious sense, though they are indeed relative to an observer and a time (and to an instrument).

Observation *statements*, on the other hand, are to be construed objectively. "The crow observed by me at 8:17 . . . *was* black" and "Sample 3 weighs between 3.50 and 3.60 grams" make direct claims about the world. As the first example of an observation report suggests, there are a lot of circumstances in which an observation report justifies the acceptance of an observation statement. Our problem is to provide an explanation of the mechanism that allows us to handle the uncertainties of observation reports, and yet use them as probabilistic grounds for accepting observation statements.

Let ML be a metalanguage for the language L. ML is to include L, as well as a vocabulary for referring to the expressions of L. In addition, it contains a recursive characterization of the *kinds* of statements that can be the subject of observational judgment. Thus "a is a crow," "a is black," and "The reading on the scale is 3.456 grams" are clearly of the right kind, while "All crows are black," "Between 85% and 95% of crows are black," and "The weight of the third sample lies between 3.50 and 3.60 grams" are not.

But there are many statements whose status in this regard is not clear: "Jupiter has three moons, " "The electron (in the cloud chamber) moved in a downward-curving path," and even "There is a plane passing overhead" (when only the contrail is visible). We shall return to the question of observation in a later chapter; for the moment, we shall merely assume that there is a class of statements in L that captures the notion of observability (perhaps relative to L itself) and that can be characterized recursively in the metalanguage.

Finally, ML contains a three-place predicate, $O(X,S,t)$, that bears the sense of: the person, group, scientific community, X, has recorded the observation embodied in S, at some time before t. The sentence S is to be a sentence of L, one of the observation statements we have characterized in the metalanguage. We specify "at some time before t" to avoid having to consider memory in detail, but it should be noted that the (relative) reliability of memory can also be handled by the methods outlined below. Of course we don't need a metric for time; it suffices that history is monotonically increasing. We shall return to this question later. Meanwhile, one may think of S in the context $O(X,S,t)$ as an observation report, written in X's laboratory notebook.

Observation reports are handy. They can be regarded as incorrigible, for example. Once written in the notebook of our experience, they need never be erased. (Indeed, that would be a scientific crime!) We may therefore consider a deductively closed metacorpus MK^* as embodying all of X's observation reports made prior to t, as well, of course, as the logical, mathematical, and linguistic truths of L, since L is a sublanguage of ML.

Now we can distinguish between veridical and erroneous observations in ML. X has judged that S veridically at some time before $t-VO(X,S,t)$ — just in case $O(X,S,t)$ and S, X has judged erroneously that S at some time before $t-EO(X,S,t)$ — just in case $O(X,S,t)$ and $\sim S$. Furthermore, though we cannot tell which of X's judgments are veridical and which are erroneous just by looking at MK^*, we can tell something about X's error rates. For example, if MK^* contains "$O(X,S,t)$" and "$O(X,R,t)$," and "$\sim (S\&R)$" is a theorem of L, then *at least* one of S and R must be wrong.

We can now apply a form of what I shall call the *minimization principle*: roughly that error is not to be imputed to X's observations gratuitously. Thus, given MK^* containing a finite number of observation reports, we attribute to those observation reports the minimum amount of error that is entailed by the axioms of L. In the example of the crows, we suppose that X has made just 10 errors: he has erroneously judged crowness between 0 and 10 times, and he has erroneously judged nonblackness between 0 and 10 times, and he *need* have made no more than 10 errors in all.[5] Just in terms of this principle, we can compute a general index of X's reliability up to time t: it is just the relative frequency of error we are *required* to impute to his judgments up to t.

On the basis of these data, we can make a statistical inference concerning X's general observational reliability. (And X is in just as good a position as we are to do this.) Probability is defined for ML as well as for L; we have all the mathematics we need in ML. Let $O(X)$ be the set of observation sentences S such that for some time t, $O(X,S,t)$ — that is, X will have accepted S on the basis of observation

---

[5]Note that there must have been at least 10 errors, but that if some observation was in error in two ways (both with regard to the bird and with regard to the color), then there must have been 11 errors.

at some time or other. As before, we may distinguish between veridical judgments, $VO(X)$, and erroneous judgments, $EO(X)$, though we may well not know which are which. In MK^* we have the minimal relative frequency of past erroneous judgments among past judgments as given by the minimization principle. Past judgments are a subset of the general class of X's observational judgments. That is, $O(X,t)$—the set of sentences S such that $O(S,X,t)$—is a subset of $O(X)$ and may—subject to the usual constraints about randomness—be a random member of the set of equinumerous such subsets with respect to reflecting the relative frequency of erroneous judgments, that is, of members of $EO(X)$ in $O(X)$.

If all this is so, it may be very probable that the relative frequency of error in the long run lies close to the observed frequency of error. Note that we do *not* have to specify which particular statements are in error. All we need for this is the relative frequency, and that we can obtain by reference to the axioms of L and the minimization principle.

But this may not be as interesting as some more specific knowledge that we can obtain. Given a language with a set of axioms, and given a set of observations $O(X,S,t)$, the minimization principle gives us the means to determine probabilistically the relative frequency of veridical observations among observations in general. A second plausible principle, the *distribution principle*, directs us to suppose that the errors of observation are as evenly distributed among *kinds* of observations as possible, where a *kind* of observation is determined by the predicate that is affirmed or denied of an individual.

For example (an example to which we will return in a later chapter), suppose that our language contains the constraint that A entails B, and that our metacorpus contains 100 observation reports of A's and 200 observation reports of non-B's, and that of these 300 observation reports we know that 10 are false, since 10 instances of A observations and non-B observations involve the same individuals.[6] The minimization principle directs that we should take the overall sample error frequency to be 3%. But we could suppose that 10 A observations are erroneous, or 10 non-B observations, or 7 A observations and 3 non-B observations, and so on. The distribution principle directs us, subject to the satisfaction of the minimization principle, to distribute the errors equally between the A observations and the non-B observa-

---
[6]We can regard the reidentification of individuals also as a particular kind of observational judgment and apply the same general procedures to get a measure of reliability for this sort of judgment, too.

tions—that is, to suppose that there are five observations of each sort that are in error. This leads to a sample frequency of error in A judgments of 5% and a sample of frequency of error in non-B judgments of 2.5%.

Again, if we were seeking an explanation, it might be that we should suppose that the error *rates* should be taken to be as nearly equal as possible. But just as we suppose no more than the minimum absolute number of errors, so we suppose that absolute number divided as equally as possible, given the satisfaction of the first constraint, among the types of errors we may make.

We should observe that in the long run—that is, with a lot of data—these differences will wash out anyway. The error rates characteristic of various kinds of observation will reflect such a large body of data that the distribution principle will have relatively little impact. Our object here is merely to demonstrate that a bootstrap procedure for getting at error rates is possible and plausible.

To continue: given the minimization principle and the distribution principle, we can thus obtain sample error rates among kinds of observations, and, again by statistical inference, infer with high probability long-run error rates among the various kinds of observations.

Suppose that we have in our metacorpus of evidential certainties (metalinguistic statements that we accept on the basis of their high probability) a set of long-run approximate error rates for the various kinds of observations. Consider an observation predicate P and an individual term i such that ⌜Pi⌝ and ⌜$\sim Pi$⌝ are basic statements in Carnap's sense, as well as potential observation statements in our framework. (Characterizing the individual terms i such that ⌜Pi⌝ is in our special class of potential observation statements promises to be considerably trickier than characterizing the predicates P, but that problem need not concern us here. And it may be that in some cases we will want ⌜Pi⌝ to be an observation sentence at the same time that ⌜$\sim Pi$⌝ is not an observation sentence.) Suppose that $O(X, \ulcorner Pi \urcorner, t)$. Three cases may arise with regard to the correct reference class for measuring the probability of error of the statement ⌜Pi⌝. ⌜Pi⌝ may be a random member of the set $O(X)$ of X's observations in general; it may be a random member of the set $O^P(X)$ of X's observations of kind P, or it may be a random member of the set of observations of kind P that conflict directly (via the axioms) with other observation reports in MK^*.

Since, as we have remarked, we always have some nontrivial information about the error rate in the set $O^P(X)$, ⌜Pi⌝ will be a random member of $O(X)$ only when the error rate in $O^p(X)$ does not differ, in our technical sense, from the error rate in $O(X)$; in general, the error rate in $O(X)$ may be more precisely known. But if ⌜Pi⌝ is one of the observation reports that conflicts with another (or a group of other) observation reports in MK^*, then, since the error rate in that subset of $O^p(X)$ is known to be greater than that in $O^p(X)$ in general, ⌜Pi⌝ will be a random member of that subset with respect to being erroneous.

There are thus (ordinarily) three reference classes that may determine the probability of error of ⌜Pi⌝:

1. If the error rate in $O^p(X)$ does not differ from the error rate in $O(X)$, then $O(X)$ provides the probability.
2. If the error rate in $O^p(X)$ does differ from the error rate in $O(X)$, then $O^p(X)$ provides the probability, since $O^p(X)$ is included in $O(X)$.
3. If ⌜Pi⌝ conflicts directly, via the axioms of the language, with some other observation report, then it belongs to a special subclass of $O^p(X)$ in which the frequency of error may be quite high — high enough, perhaps, to preclude our even accepting ⌜Pi⌝.

The result of observation is ordinarily that when "$O(X,$ ⌜Pi⌝$,t)$" is in MK^*, the probability (relative to the metacorpus of evidential certainties, MK) of ⌜Pi⌝ is very high unless ⌜Pi⌝ is one of those observations that conflicts directly with other observations. In general, then, the observation report that Pi warrants the acceptance of ⌜Pi⌝ in the corpus of evidential certainties. We can acknowledge the existence of observational error in general, and even be unable to specify any particular observation statement that is certain, and yet be able to accept most observation statements on the basis of the corresponding reports and to reject others.

This is true of categorical qualitative observation reports. Measurement and quantitative assertions are another matter, to which we will turn in the next chapter. The crucial question, in categorical or quantitative observation reports, is the reference class to which that report is to be referred for the purpose of assessing its reliability. On the interpretation of probability we have been assuming all along, that question always has an answer. Given an observation report, there is some minimum probability, determined by the right reference class for the reliability of that particular report, that it (or, in the case of quantitative statements, a related interval statement) is true.

6

Measurement

Quantities

It is easy to think of dynamics as the paradigm of a sophisticated scientific theory. Quantitative laws, it seems, are what real science is really about. Thus we see the social sciences — psychology, economics, even some parts of anthropology and history — striving to achieve quantitative status. And yet, many of the most impressive scientific achievements of the past, like many of those of recent years, do not appear to be quantitative. The atomic theory, for example, though it was inspired by and accounted for the constancy of combining ratios for chemical reactions, was essentially a qualitative structural hypothesis about the nature of matter. Many of the major achievements of recent years concern the uncovering of the structures of particular (biologically interesting) molecules. The basic idea of evolutionary theory is essentially qualitative in character.

Nevertheless, there is more to the claim that modern science is essentially quantitative than these examples might suggest. The data provided by constant combining ratios depend crucially on a quantitative view of chemical reactions and on measurements that were, for the time, extremely precise. Molecular structures are inferred from quantitative data. Even evolutionary theory depends both on quantitative data concerning geological events and, more recently, on quantitative data concerning differential rates of reproduction. Quantitative data lie behind (sometimes rather far behind, it must be admitted) most of those disciplines that lay claim to being scientific.

There is a cheap version of this argument that is worth considering. Suppose that P is a qualitative predicate and i is an instance to which it might apply: Pi is either true or false. We can always replace the qualitative predicate P by the quantitative predicate P^*, applicable to the same class of things, that takes two values: P^*i has the value 1.0 if Pi is true and the value 0.0 otherwise.

On the surface, this is a deceptive and factitious change. But it can have a point. The expected value of P^* over a population of individuals to which P might apply is just the relative frequency with which individuals in that population have the property P. An estimate of the former is an estimate of the latter. Furthermore, given a single instance i, the expected value of P^*i taken over a number of individuals who make judgments about whether i is P, that is, whether $P^*i=1$, is just the relative frequency with which their judgment is positive. We can thus use the expected value of P^*i is a measure of the vagueness or fuzziness[1] of the predicate P. It can also be used as an indication of the reliability with which P is used by that group of people.

I am not sure how to use this bit of machinery in the former way; it is unclear what specific values of P^*i, other than 0.0 and 1.0, really mean. Can something be nine-tenths a cow? But the latter use has a natural value. It can enter into the quantitative treatment of error of judgments concerning the qualitative predicate P, provided that we have some independent hold on the truth.

The general claim, though, that quantitative data are the underpinning of the most qualitative of modern scientific theories seems sound. It is quite clear in the cases of biochemistry, cell biology, and other disciplines that make material use of such clearly quantitative theories as those of physics and chemistry. It is somewhat less clear in the case of such disciplines as sociology and epidemiology, where one is often essentially counting cases and subjecting the results to statistical analysis. But even then, quantities, sometimes crude, such as years of schooling, income, number of siblings, and so on are involved.

We had best, therefore, explore the controversial question of what quantities are. It should be noted that in statistics quantities are often called "random variables" — an unwarranted and misleading practice, since so-called random variables are functions, and variables are linguistic entities. Some writers in statistics refer to these objects as "random quantities," but the adjective is merely being used to call attention to the assumed stochastic nature of the underlying process. In any

---

[1] Zadeh has made the most of this approach. See, for example, "Fuzzy Logic and Approximate Reasoning," *Synthese* **30**, 1975, 407–428.

event, a quantity (random quantity, random variable) is a function applicable to a certain class of objects. Thus "length of" is a function whose domain is material bodies; "distance" is a function whose domain is pairs of points in some metric space; "intelligence quotient" a function whose domain is people; "weight" is a function whose domain is ponderable bodies near the earth's surface. This much is relatively uncontroversial.

Controversy begins when we consider the ranges of the functions that are quantities. Many writers—the most precise and thoughtful being Carnap[2]—take the range of a quantity function to be the real numbers. This is awkward because it requires us to speak of a number of different length functions: the length in feet of function, the length in meters of function, and so on, and we must then account for the conventional relations among these functions. It has the advantage that we do not have to introduce *magnitudes* into our scientific language, where a magnitude is an abstract entity that is the value of a quantity function applied to a particular object. A magnitude is what objects with the same length have in common. It is alleged that dispensing with magnitudes is desirable, since for each kind of magnitude (e.g., length, mass, amperage, . . .) we will have to suppose the existence of a whole nondenumerable infinity of abstract objects that we need otherwise not postulate. Ontological caution suggests that we not introduce more objects—particularly abstract ones—than we need to introduce.

The counterargument, which I find more persuasive, is that by eschewing magnitudes, we fall into an ontological inflation of functions. There are any number of possible units (e.g., feet, meters, miles . . .) that we can choose for the measurement of length. One might claim, as persuasively as one can claim that one needs a nondenumerable infinity of length magnitudes, that there is a nondenumerable number of length functions of the form length-in-*blank*-of, where *blank* is replaced by a chosen unit of length. There thus seems to be no motive after all in departing from the conventional scientific procedure of including reference to the unit as part of the *value* of the function: we report "the length of rod A is 31.45 centimeters." Carnap's alternative would have us say, strictly: "the length-in-centimeters-of A is the real number 31.4999. . . . "

[2]Rudolf Carnap, *Philosophical Foundations of Physics*, Basic Books, New York, 1966. See also Ernest Nagel, *The Structure of Science*, Harcourt, Brace and World, New York, 1961.

Furthermore, while magnitudes are admittedly abstract, it is not clear that we need a nondenumerable number of each kind, and it does seem that they aren't *very* abstract. (Surely they are no more abstract than numbers!) For, following a Russellian treatment of natural numbers, we can define a magnitude of a certain kind (length, for example) as the equivalence class of objects standing in a certain relation. Thus the length of A, the magnitude that is A's length, is the set of objects equal in length to A. This has the advantage that the magnitude of length denoted by "100 centimeters" is, strictly speaking, *identical* to the magnitude denoted by "1 meter," and even that the magnitude denoted by "2.54005 centimeters" is the same as the magnitude denoted by "1 inch." It also has the advantage that the cardinality of the set of magnitudes of a certain sort cannot exceed the cardinality of the domain of the corresponding quantity, which may sometimes even be finite. Of course, that also means that not every length expression need denote: there may be no magnitude denoted by 1.999 . . . centimeters, and none denoted by 2.999 . . . centimeters, and so both of these magnitudes may turn out to be the same, namely, the empty set. But since we could never know such a thing, it doesn't seem necessary to worry about it.

Treating quantities in this way has other advantages. It makes sense, for example, of the role of dimensional analysis in science and engineering. It makes sense of the fact that many quantities do not have to use many of the real numbers available: objects don't have negative masses, for example, nor do people have irrational numbers of siblings, even though they may have numbers of irrational siblings. Furthermore it leads us to focus on interesting questions in the philosophy of science. Thus it is ordinarily supposed that absolute temperature can only have values corresponding to the nonnegative real numbers. It has recently been argued[3] that negative and hotter than infinite temperatures make sense. This argument is not about the range of the temperature-in-degrees-Kelvin-of function (there is nothing unnatural about negative real numbers!) but about the relation between magnitudes of temperature and the rest of physical theory. When we think of magnitudes, it is easy to have our attention directed to the theories in

---

[3] Philip Ehrlich, "Negative, Infinite, and Hotter Than Infinite Temperatures," Manuscript, 1980.

which the magnitudes play a role, and to the interaction between theory and measurement.[4]

Measurement and Observation

Given that quantities are functions from objects (or processes or whatever) to magnitudes, how do we go about measuring them? There is one easy way: since the length of an object is the set of objects with the same length as it, we can adopt that object as the unit of length and say that it is 1 unit long. Exactly! To any number of decimal places! But that is not very helpful, since the reason we want to measure is to establish relations — and especially easily communicable relations — among objects.

Basic measurement begins when we have an observable, transitive (for the most part) relation among objects. Some rigid bodies are longer than others, and some processes endure longer than others. Furthermore, if A is longer than B and B is longer than C, in general A will be found to be longer than C. Note that the converse relation tends *not* to be uniformly transitive: from A is not longer than B and B is not longer than C, we can conclude only insecurely that A is not longer than C. This is because A and B, and B and C, may differ only subliminally in length, while A and C differ noticeably.

If we remove ourselves a little from the constraints of observability, however, we are free to suppose that "not longer than" is also transitive. In the next section and in the next two chapters, we will consider the grounds for making such a supposition. If "not longer than" is taken to be transitive, then we may define "is the same length as" simply as: A is the same length as B just in case A is not longer than B and B is not longer than A. This is an equivalence relation (given our supposition), and thus we can define equivalence classes of lengthy objects or, in general, objects in the field of whatever transitive relation we started with.

So far, no numbers and no scales. But we have introduced magnitudes. We also have an empirical way of telling when the magnitude possessed by one object is greater than that possessed by another — not

---

[4]For an exploration of this claim, see my *Theory and Measurement*, Cambridge University Press, Cambridge, 1984.

without error, to be sure, but one that is often relatively reliable. We look to see when the one object bears the transitive relation in question to the other. In order to introduce numbers and scales, we have to introduce a *unit* and a way of relating objects in the field of the basic relation to the standard established by the unit. It is important to note that the *unit* need not be represented by a single object—for example, the tired old platinum meter stick outside of Paris. In the case of temperature, for example, the *unit* may consist of *two* standard objects: a mixture of ice and water (assigned the standard temperature 32° Fahrenheit) and a mixture of water and steam (assigned the standard temperature 212° Fahrenheit).

The unit chosen for a magnitude, together with the number or numbers representing the conventional points of the unit, do not determine the whole scale. We must also have a way of relating other objects in the field of the relation to the basic unit. In the case of temperature, we might take *mixture* as the fundamental procedure: if something is the same temperature as a mixture of equal parts water at 0° (centigrade) and water at 100° (centrigrade) then we may assign it a temperature of 50° centigrade. In the case of the most useful and fundamental quantities, we can do better. These quantities (length, distance, mass, time, etc.) admit of scales that are called "ratio scales." What is special about these scales is that they require only a single unit (for distance, it used to be that platinum meter stick at Sèvres), and that any two scales for these quantities are related by a multiplicative constant.

Let us first consider the selection of a unit. In the classical case of the selection of a unit of length, the meter was *defined* as the distance between two fine marks on a certain physical object at a certain place. It is sometimes suggested that this is typical, and that it is the particular physical object in question that constitutes the unit. This leads to the somewhat frivolous question of whether, if that object shrank, we would not rather be obliged to say that everything else in the whole universe expanded. But it makes more sense to say that the unit is not the segment of the platinum bar between the two marks: it is the *length* of the segment of the rod between the two marks, and this length is a *magnitude*—an abstract object consisting of the set of all objects (in the whole universe!) 1 meter long, together with (for we might as well lump distances and lengths together explicitly) all pairs of points 1 meter apart. The function of the particular bar of platinum referred to as the "standard" was merely to help us pick out this class.

Physical objects can change in length. The boiling point and freezing point of water vary with pressure. A large rubber band can be 1 meter long at one time and less than 1 meter long later. Over time, objects can become members of, and cease to be members of, the equivalence class of 1-meter objects. (Strictly speaking, therefore, it is objects-at-times or objects-during-periods that belong to these equivalence classes.) This is as true of the canonical meter standard as it is of any other object. We take steps — those we know how to take — to avoid this change in status: we keep the standard meter bar at 0°C; we support it in a special way so that the stresses induced by gravity will not induce strains that change its length. But despite our best efforts, it may be the case that even the standard meter bar will cease to be a member of the equivalence class of objects 1 meter long. (An angry government employee might attempt to destroy the whole edifice of physics by hitting the bar with a sledge hammer!)

To characterize the unit of length, what we need is something stable. For many years the platinum bar, carefully preserved under standard conditions, served that function. But we have since discovered a standard that is both more portable (only *copies* of the standard bar could be carried around) and less subject to environmental influence: the wavelength of light emitted by the cesium atom in a certain state of excitation. This standard is not accessible to the seller of dry goods, but neither was the platinum bar. What is important is that there exists a court of last resort — a way of *specifying* the unit of length.

There are many ways of specifying the units of the various quantities involved in science. (The ampere of current, for example, is specified in terms of the mass of silver deposited on an electrode under specified conditions.) Specifying the unit, however, does not determine the *scale* on which a quantity is to be measured, though in fact there is a close connection between the form of the unit and the scale. Length, mass, and time, for example, are measured on *ratio scales*. This is possible because these quantities are additive: there is a natural and convenient way of combining objects to which the quantities apply that can be used to generate the scale (collinear juxtaposition in the case of length, for example). Whatever unit we choose for length, the length of the collinear juxtaposition of two bodies is the sum of the lengths of the two bodies. From this it follows that the lengths on one scale are a constant factor times the lengths on another scale.

But wait! We haven't introduced *numbers* at all.[5] What is the relation between numbers and magnitudes? On the view that takes the range of quantitative functions to *be* numbers, the relation is just one of identity. (And then there is no reason at all not to add temperatures to time intervals!) On the present view, matters are a bit more complicated. A set of magnitudes has some of the structure of a set of numbers. Distance has all of the structure of the reals, and mass has the structure of the nonnegative reals. But even quantities in the weakest sense have some of this structure: as soon as we have equivalence classes under the converse of the original asymmetric relation, we have an ordering, and this can be reflected in the order properties of numbers.

Every magnitude can be represented as a real number (or vector) times the unit of that magnitude in such a way as to reflect the structure of that set of magnitudes. It is magnitudes expressed in this canonical form that we have in mind when we speak of the value of a quantity. Furthermore, magnitudes expressed in a common unit can be added (note that we can add temperatures, even though temperature is not "additive"; in fact, we *must* do this to compute an average temperature) and can be multiplied by a scalar constant. In short, magnitudes form a vector algebra, except for the fact that it is not always the case that there is a magnitude corresponding to each real number times the unit. Nothing has a mass of -7 grams. And when hardness is measured on the mho scale of 10, a hardness of 4.3756 does not determine a magnitude of hardness at all.

To sum up: the observational basis of any quantitative measure is a transitive, asymmetric relation (longer than, hotter than, . . .). It is (creatively and imaginatively) assumed that the converse relation (not longer than, not hotter than) is also transitive, and therefore can be used to generate equivalence classes (is as long as, is as hot as). A particular equivalence class or set of equivalence classes (a particular object, a particular pair of temperatures, . . .) is selected to serve as a unit (to determine a scale) and is assigned a particular numerical value or set of values (1 meter, 100°C, and 0°C). A particular procedure (collinear juxtaposition, mixing of liquids) is adopted to relate the nu-

---

[5]In a sense, we don't have to. This is the thesis of Hartry Field's book, *Science Without Numbers*, Princeton University Press, Princeton, N.J., 1980.

merical value of an arbitrary magnitude (the length of this rod, the temperature of this stuff) to that (or those) of the unit.[6]

So far, we have only addressed the question of *direct* measurement. Since all measurement ultimately reduces to direct measurement, what we have already said will suffice to establish a framework for talking about errors of measurement in general. We shall return later to the question of indirect measurement.

Measurement and Error

It is all very well to define magnitudes, but that does us no good unless we can actually measure them. It is true that we can measure some quantities, namely, the uninteresting quantities provided by the unit. We know that the meter bar in Sèvres is 1 meter long, that under standard conditions the melting and boiling points of water are 100°C apart, and that the length of an arbitrary rigid body is exactly 1 unit long, where the length of that body is taken to define the unit. But none of this does us much good.

In order to measure things in terms of a *common* unit, which is exactly what we must do in order to communicate with each other about magnitudes and in order to compare objects that are separated in time or space, we must introduce the notion of *error*. It is perhaps ironic, but nevertheless true, that in order to achieve precision and exactitude, we must introduce error and approximation.

One difficulty in attempting to elucidate how quantitative exactitude and error go hand in hand is that in both human history and in the history of each individual, various quantities are highly correlated with other quantities. Everybody knows that in general bigger things weigh more than smaller ones and take longer to move from here to there. So in order to illustrate how error and measurement can be developed from scratch, we must imagine ourselves in an extremely artificial hypothetical situation.

With that warning, let us look at length. We take for granted that we can distinguish, with a high degree of certainty, (relatively) rigid bodies. We take for granted that we can tell a collinear juxtaposition when we see one with a high degree of certainty. And we take for

---

[6]The details of such procedures are worked out in *Theory and Measurement*.

granted that we can tell with a high degree of certainty when one rigid body is longer than another. It does not take a great leap of faith (or an excess of caution in making judgments) to suppose that *being longer than* is transitive. This is as far as observation alone can take us.

We now suppose that *not being longer than* is also transitive. This supposition is gratuitous and theoretical. We are already faced with judgmental counterexamples: I cannot see that A is longer than B; I cannot see that B is longer than C; but I can see that A is longer than C. I suspend my judgment, and perhaps my common sense, and declare that one of these judgments—A is not longer than B, B is not longer than C, and A is longer than C—embodies an error. But, as we shall see, this assumption of error can pay off.

Given that the relation *is not longer than* is transitive, we can define equivalence classes under the relation *is the same length as*. (Recall that we define "A is the same length as B" as "A is not longer than B and B is not longer than A.") We now have, on purely logical grounds, *magnitudes* of length—that is, equivalence classes of length. But we do not have numerical lengths. To get numerical lengths, we must pick a unit, and pick a procedure for relating new lengths to old ones or for generating arbitrary lengths. Let us pick a unit and assign it the value 1.0 meter, and let us use collinear juxtaposition as our method of extending measures of length to new objects. This presupposes the additivity of length, but we will consider the grounds for making such an assumption in a later chapter. We can then reconstrue our meter standard as the collinear juxtaposition of 1000 bodies 1 millimeter in length. We can even mark it, thus obtaining the standard meter stick with which we are all familiar.

Now let us measure a rigid body repeatedly, recording our results (in millimeters) in our notebooks (or our metacorpora **MK***). We obtain a certain distribution of numbers: 580.0, 587.6, 587.5, 588.2, and so on. From this sample distribution, we can infer statistically that it is highly probable that the population of measurements of which these constitute a sample is distributed in a kind of bell-shaped way, with a mean between (say) 588.1 and 588.3 and a variance of approximately 1.1.

The length of the object we are measuring is a certain magnitude. Every magnitude can be represented as a constant times our defined millimeter magnitude; 588.1 millimeters is a constant times a millimeter magnitude, but how is it related to the magnitude of the object we are measuring?

More generally, what is the relation between the set of numbers we obtain, each of which reflects a certain amount of error, and the true length of the object? Provided that all the information we have is the sample of measurements, we may apply the minimization principle: take the "true value" to be that value that minimizes the errors of observation. It is clear that the mean of the observed values minimizes the errors, as measured by the square of the difference between the observation and the true value.[7]

From measurements of a single rigid body, together with the minimization principle, we obtain a *sample distribution* of errors of measurement. This sample distribution is represented in the deductively closed metacorpus **MK***: the notebook in which our observations are recorded. By statistical inference, then, it can be highly probable, relative to **MK***, that the general distribution of errors, of which we have an observed sample, is one of a certain family of distributions. The distributions in this family have means close to 0 and variances close to that in the sample; and in fact, if we have enough data, these distributions will be close to normal.

But our measurements need not be all of a single body. Suppose that we have made measurements of a number of rigid bodies. Applying the minimization principle to each set of measurements independently, we obtain a collection of samples of error distributions. Can we lump these together to form a larger sample on the basis of which we can infer with high probability a more exact distribution of error? Ignore, for the moment, relations among the lengths of the rigid bodies we have measured. Then the answer to this question is given by the internal structure of our collection of samples. It may be that the large sample can be regarded as a random member of the set of equinumerous samples of error distributions with respect to reflecting the general distribution of errors of measurement of the sort we are contemplating. Or it may be that certain sets of measurements—for example, those of very large objects or those of very small objects—must be treated separately. What is at issue here is the question of what statistical inference is warranted by the data. Assuming, as we are, that statistical inference can be made sense of (e.g., in roughly the way outlined in *The Foundations of Statistical Inference*), we can often combine our data to get large samples on the basis of which we can

---

[7]Minimizing the mean squared error is analytically simpler than minimizing the mean absolute error.

probabilistically infer quite precise general error distributions. Put more exactly: we are often *not prevented* from thus combining our data by their internal structure.

Now let us take account of relations among the objects being measured. The only relation we need worry about at the moment is that expressed by the additivity of length. We know (because we decided that this was to be a property of length) that the length of the collinear juxtaposition of x and y is to be the sum of the lengths of x and of y, where the sum of k feet and j feet is to be $(k+j)$ feet. This imposes an *additional* constraint on our observations. To reconcile our observations to this new constraint requires that we recognize more error in our observations. For example, suppose that the measurements of x yield 3, 4, and 5 feet, measurements of y yield 7, 8, and 9 feet, and measurements of their collinear juxtaposition yield 13, 14, and 15 feet. Without the additivity law, we need merely suppose that we have made six errors of magnitude 1 foot.

With the additivity law, we must suppose that we have made more error. How much? We must bring in the distribution principle: assume that the errors are distributed equally among the three classes of measurements. We can see that the minimum total error is 8 feet (take the length of x to be 5 feet and the length of y to be 9 feet). Divided among the three classes, this is 2⅔ feet in each, so the values of the three lengths that satisfy the additivity law *and* minimize the error *and* yield uniformly distributed error are 4⅔ feet, 8⅔ feet, and 13⅓ feet.[8] Note that the *mean* error in each class of measurements is no longer 0.

It might turn out that the errors we discovered were systematic — for example, that things tended to shrink into each other when juxtaposed or to expand on contact. But a "systematic" error, on the interpretation of probability offered here, is just a random error with a different reference class: we would just say that measurements of juxtaposed bodies are subject to a different distribution of error than are measurements of other bodies. (In reality, of course, we would take this as evidence that it is ill-advised to regard length as additive. This use of *evidence* will be considered in detail later.)

In fact, no such thing happens, and the added constraint that length is additive has a very small effect on the inferred error distribu-

---

[8] More formally, take the three lengths to be t_1, t_2, and t_1+t_2. Compute the sum of the squares of the errors of the measurements (in terms of t_1 and t_2), getting
$$SE = 6(t_1^2 + t_2^2 + t_1 t_2 - 18 t_1 - 22 t_2 + N)$$
where N is constant. Minimizing the squared error SE yields $t_1 = 4⅔$ and $t_2 = 8⅔$.

tions, though since we are satisfying an additional constraint, the effect will not be zero.

It is often the case that an observable transitive relation can be made to yield a quantity. These are the quantities that admit of *direct* measurement. They include many quantities (temperature, area, velocity, . . .) that one does not think of in terms of direct measurement, because for these quantities *indirect* measurement—measurement in terms of some other quantity, as a mercury thermometer provides a measurement of temperature in terms of the length of its mercury column—proves to be more accurate and more convenient.

More *accurate*? Yes; given laws in our language connecting the quantities involved, we can approach indirect measurement in the same way that we approached length under the assumption that it was additive. We may discover that we need to assume less error in the case of indirect measurement than in the case of direct measurement of certain quantities.[9]

Indirect Measurement

Most quantities with which science is concerned are measured indirectly, even when they do admit of direct measurement. By "direct measurement" I mean the judgmental comparison of the object being measured with another object in the respect that is being measured. Thus measurement with a micrometer is indirect (since one is using length to measure angle to measure length), while a measurement of length under a microscope of one item against another of known length is direct. There are some quantities (charge, voltage, intelligence, etc.) that can *only* be measured indirectly. There are generally various ways in which a given quantity can be measured indirectly.

Temperature, for example, is measured indirectly in terms of length when we use a mercury thermometer; it is measured indirectly in terms of angle when we use a bimetallic strip connected to a spring and a needle; it is measured in terms of volume when we use a gas thermometer; it is measured in terms of the on/off characteristics of light-emitting diodes when we use an electronic thermometer; and in the case of very high temperatures, it is measured in terms of the colors of standard substances ("cherry red").

---

[9]For a relatively detailed discussion of these matters, see *Theory and Measurement*.

In every case of indirect measurement, though, we depend on some quantitative relation between the quantity being measured indirectly and the quantity in terms of which that measurement is made. It has been suggested[10] that this quantitative relation can be taken as *defining* the quantity measured indirectly; thus the length of the mercury column can be taken as an "operational definition" of temperature. There are several reasons for rejecting this suggestion, and indeed for taking a skeptical attitude toward operational definitions in general.

Most quantities, even those that are almost always measured indirectly, admit also of direct measurement and are such that we can sometimes make direct judgments of inequalities. These direct judgments and direct measurements must in general (not necessarily always) be taken as norms to which indirect measurements must conform. The indirect measurement may be taken to be "more accurate" than the direct measurement—in ways we shall describe shortly—but it cannot have the result that the direct measurement is *wholly* inaccurate or that our direct comparative judgements are *generally* in error. Thus operational definitions cannot be taken as *definitions*.

In general (temperature again provides a fine example), indirect measurements of different kinds are appropriate at different regions of the scale. Thus we use a mercury thermometer at certain ranges of temperature, a gas thermometer for higher ranges, an alcohol thermometer for lower ranges, a thermocouple for certain other ranges, and so on. There need be no *single* procedure for the indirect measurement of a quantity. But these ranges overlap extensively, and in the regions of overlap it is arbitrary (and unnecessary) to choose one particular method of indirect measurement as the definatory one.

Furthermore, these procedures of indirect measurement are more accurate in some regions of the scale than in others. One can hardly speak of a "definition" as being more or less accurate. And in fact, we can conduct indirect measurement perfectly well even when we know that the quantitative relation on which it is based is inaccurate. A mercury thermometer, for example, is constructed on the supposition that the expansion of mercury is a linear function of its increase in temperature. In fact, we know that that is not strictly true (see any handbook of chemistry or physics). Nevertheless, (a) it is true enough for most purposes, and (b) if it isn't true enough for *our* purposes, we

---

[10] See P. W. Bridgman, *The Logic of Modern Physics*, Macmillan, New York, 1927.

introduce correction factors or calibrate the instrument against a standard.

Even though we cannot *define* a quantity by reference to indirect means of measuring it, it may well be that indirect measurements far surpass direct measurements of the quantity in accuracy as well as ease. What can this mean? How can an indirect measurement be more accurate than a direct measurement? In the discussion of the direct measurement of length, we noted that it was possible that the imposition of the constraint that length be additive could have led to systematic errors of measurement. Similarly, the imposition of constraints on other quantities may well require that we regard their direct measurement as subject to systematic errors. Furthermore, the quantities we are most prone to measure indirectly are those whose direct measurement is particularly prone to random error. An indirect method of measurement may decrease "random" error, in the sense that results of the indirect measurement may have a far smaller variance — the results are more "consistent" — than the results of direct measurements and do not conflict with them. We can also give an appropriate sense to "systematic error" in which indirect measurement may eliminate it: namely, when a number of methods of indirect measurement cohere together better than any one of them does with the method of direct measurement, we sensibly conclude that it is the direct method that embodies the systematic error.

Finally, different kinds of indirect measurement may be accurate in different regions of the scale. All of this, of course, requires that we know how the quantities we measure directly are related to those that we thereby measure indirectly. How we know such things will be the subject of the next few chapters. It suffices here to note that we can know them without demanding that they be definitional truths.

We can construct a rather crude example by considering temperature. Taking the *unit* of temperature to be the boiling and melting points of water, we can measure temperature directly, if crudely, by constructing a standard scale in terms of mixtures of determinate quantities of water (or any other liquid) and making comparative judgments. A body that is neither hotter nor colder than a mixture of equal parts of a liquid the same temperature as an ice-water mixture, and the same liquid at the temperature of boiling water, is at 50° on the Celsius scale.

Our comparative judgments aren't very precise here, but they are precise enough for us to formulate and test the law of linear thermal expansion. Since we can measure distances very well, this law gives us a way of measuring temperature that is internally very accurate. Is it *really* very accurate? So long as we have just this one way of measuring temperature indirectly, and it does not conflict with our direct measurements, we have no grounds for doubting its accuracy.

But taking the law at face value and using it to measure temperatures, we can observe that not all substances obey the law equally well at all temperatures. This can lead us to refine the law. Add to this the fact that we can, in the same way, develop other ways of measuring temperature indirectly, in terms of other empirical laws, and the whole process may lead to a collection of indirect methods of measuring temperatures, none of which need be taken as perfectly accurate over any significant range.

Ultimately, we can characterize temperature in terms of mean kinetic energy. In a rich theory, we *can* "define" temperature this way. But that is because the identification of temperature with mean kinetic energy, in the presence of all the knowledge we have of physics, yields an account of where, how, and why our accurate indirect measurements of temperature are accurate, and of where, how, and why failures of accuracy occur. We obtain, simultaneously, a theoretical account of temperature and a theoretical account of temperature measurement.

Throughout this whole process, the *random* errors of indirect measurement can be derived from the quantitative relation between the quantities directly measured and the quantity being indirectly measured. By "random," of course, I mean those whose chances (relative to what we know) are given by frequencies among the kinds of direct measurement involved in making the indirect measurement.

If X is measured as $aY+b$, the variance of the indirect measurements of X will be a^2 times the variance of the direct measurements of Y; if X is measured as $Y+Z$, and errors in the measurements of Y and Z are taken to be independent, then the variance of the indirect measurements of X will be the sum of the variances of the direct measurements of Y and of Z; and so on. And according to the view of probability we are working from, to lack knowledge of the dependence of the errors of measurements of Y and of Z does warrant their treatment as independent. This is because a pair of Y,Z observations, when we do not know otherwise, is a random member of the cross-product of the set of Y observations and the set of Z observations, and the error distribution on this cross-product is just the product of the error distribution for Y observations and the error distribution for Z observations. To know something about the dependence of the Y and Z errors is exactly to know that we are in some subset of this cross-product or in some other reference class altogether.

The derivation of error distributions in the case of indirectly measured quantities may not always be easy—particularly if there are interactions to be taken account of—but in principle it is straightforward. It generally rests on the error distributions we know to apply to quantities that we measure directly. These error distributions, in turn, are obtainable directly by means of statistical inference applied to the set of observation reports captured in the basic metacorpus **MK***.

7

Choosing Among Conventions

Conventions

Poincaré[1] maintained that the choice of a geometry in which to do physics was merely conventional, that there was "no truth of the matter" as to the geometrical structure of space. By this is meant that there are no facts, no truths about the world, that can make it true or false, for example, that the sum of the interior angles of a triangle is 180 degrees. To examine triangles composed of wooden sticks or of light rays can help only if we can believe that their edges are "straight lines." But we have no criterion of straightness other than conformity to one geometry or another. Thus we cannot refute Euclidean geometry by measuring angles. We can therefore only judge one geometry to be more convenient than another. Furthermore, even in the face of Einstein's system employing non-Euclidean geometry, Poincaré maintained that Euclidean geometry would ultimately always be preferable in virtue of its simplicity.

On this last point, most people acknowledge that Poincaré was wrong: what counts for us is not the simplicity of the geometry alone but the simplicity of the whole system of geometry-plus-physics. On the first point, that you can translate a system employing one geometry into a system employing another and vice versa, Poincaré was conclusively right.

If we *were* merely concerned with geometry, then it seems that Poincaré would be right in principle: no fact of the matter would determine one geometry to be right and another wrong; no set of

---

[1]Henri Poincaré, *Science and Hypothesis*, Dover, New York, 1952; first published as *La science et l'hypothèse*, Paris, 1902.

observations could confirm or support or render probable one geometry at the expense of the other. So far, the choice of a geometry would be conventional. But it is not conventional in the very strong sense that a different convention could *just as well* have been chosen. Poincaré specifically mentions *simplicity* as a ground for choosing between geometrical conventions, but other nonepistemic considerations might enter in as well: familiarity, elegance, ease of application. All of these properties could stand clarification. "Simplicity," for example, has generated a whole philosophical literature and is still vague enough to cover a multitude of motivations for preferring one system to another.[2]

As soon as we expand our concerns beyond geometry to, say, geometry-plus-physics, matters become even more difficult to sort out. As Duhem points out at length,[3] the idea that we can confront physical theory with experimental tests is a myth. The simplest physical experiment involves a vast number and a wide variety of fragments of physical theory: the theory involved in the design and operation of the measuring apparatus that both tells us what the initial conditions are and tells us what the outcome is. Logically speaking, it is the system as a whole that is put to the test, and not that particular fragment we are questioning, even though we may have already decided what particular proposition we will reject if the experiment does not conform to our prediction. But in any event, it will *not* be the laws on which our measurements depend that will be rejected: those are taken to be irrefutable in this context.

More than this, however, we find in Duhem an anticipation of Quine's "underdetermination" thesis: the experiments may establish experimental facts, but to every experimental fact there corresponds an infinity of possible theoretical facts.[4]

Generalizing, we are lead to Quine and Ullian's view[5] that our scientific knowledge is an interconnected web that is tied to experience

---

[2]See, for example, Patrick Suppes, "Nelson Goodman on the Concept of Logical Simplicity," *Philosophy of Science* **23**, 1956, 153–159; and Lars Svenonius, "Definability and Simplicity," *Journal of Symbolic Logic* **20**, 1955, 235–250.

[3]Pierre Duhem, *The Aim and Structure of Physical Theory*, Princeton University Press, Princeton, N.J., 1954. (First published as *La theorie physique: son object, sa structure*, Paris, 1906.)

[4]Ibid., p. 19.

[5]W. V. O. Quine and J. S. Ullian, *The Web of Belief*, Random House, New York, 1970. The "web" was a "man-made fabric" in W. V. O. Quine, "Two Dogmas of Empiricism," *Philosophical Review* **60**, 1951, 20–43.

only at the edges, where observation statements occur. Observation statements are infallible,[6] so a displacement at the edge, imposed by observation, is transmitted to the interior of the web. But it is transmitted only as a strain that may propagate in any number of ways. Again, simplicity, familiarity, and other nonepistemic considerations determine how we will actually make adjustments in the interior that will bring the edges into conformity with what happens to us.[7]

Furthermore, even if the edges—the whole set of observation statements—were completely fixed and determined, there would still be any number of ways of designing the interior system to fit those edges. This is the thesis of underdetermination: even the totality of possible experience does not determine the structure of our theory.

Finally, as Craig's theorem shows,[8] we can erase the interior of the web, leaving only the observation vocabulary and the observation statements, together with classical logical apparatus. As a trivial corollary, we can erase the interior of the web and replace it with whatever structure and theoretical terms strike our fancy, as I have noted elsewhere.[9]

All of this smacks of philosophical and logical trickery. Leaving to one side the notorious difficulties of deciding what should count as observational vocabulary and observational statements, it is hard not to feel that there is a difference in *epistemic* status between Einsteinian and Newtonian mechanics, despite the fact that, in principle, either can be squared with our observations (by the stipulation of distorting universal forces, for example). And it is hard not to feel that this epistemic difference is formal—that it would be possible to capture it in general formal constraints.

The framework offered in the earlier chapters embodies the conventionalism that is associated with Poincaré, Duhem, and Quine with a vengeance. It was suggested that adopting one theory rather than another is adopting one way of talking rather than another. Even *probability* is relativized to a language, so that comparisons of probabilities across languages can be uninformative.

[6] Or so it appears in "Empirical Content," in W. V. O. Quine, *Theories and Things*, Harvard University Press, Cambridge, Mass., 1981, pp. 24–30.

[7] But there will be "ties for first place"; W. V. O. Quine, *Word and Object*, John Wiley, New York, 1960.

[8] W. Craig, "Replacement of Auxiliary Expressions," *Philosophical Review* 65, 1956, 38–55.

[9] In "How to Make Up a Theory," *Philosophical Review* 87, 1978, 84–87.

One might go even further: why exclude the observational vocabulary? An observation sentence may be invoked by some particular pattern of irritations of the sensory apparatus, but it does not depend for its truth (or, more to the point, acceptability) on any *particular* set of irritations. Should we take the set of observation sentences to be underdetermined by the course of our sensory experience? But I find Quine's talk of stimulations and irritations less than helpful. We do not *observe* these stimulations and irritations, except under very special introspective circumstances. We observe the world around us. We use a certain vocabulary to record our observations. According to the formal semantics of our language, this vocabulary serves to partition the world: there is a certain class of things in the world satisfying the predicate "is red."

Our language does not come with a formal semantics, however. It comes with a collection of usages, and it is these that enable our observational vocabulary to slice up experience in a certain coarse way. With the very same observational vocabulary, we could slice up experience in another way by locating our errors of judgment in different places. What our observation predicates *mean* is not determined by our dispositions: we might say that there is no fact of the matter about whether *a* is red or *b* is blue when both judgments have been made and both cannot be true. No observation statement is immune from challenge by other observation statements, and no observation statement is refutable by any collection of observation statements. Even the edges of Quine's web should not regarded as immune from conventional revision.

There is no class of things in the (real) world satisfying the predicate "is red"; to suppose that there is is to suppose that there is a truth of the matter in all cases about whether a given object is red or not. But not all who speak English have the same dispositions with regard to redness; the language is alive; meanings change; there is often no unique way to assign errors of observation (though our rules do give us ways to *apportion* them); and the course of our collective experience may well have an effect on how we use our basic observational vocabulary to slice up our experience.

Of course, there is a chicken-and-egg aspect to this. We have certain innate propensities to notice and to group together certain parts of the world, and this bears on the question of what sorts of language it is feasible to learn quickly, especially when one is rather young. At the same time, the language we know and love influences how we *perceive* the world. But why should not our ways of perceiving and our ways of talking develop together? And indeed, develop from innate propensities that may be more or less influential, without being decisive, in that development?

But now I want to ring a change on the conventionalist theme. We have already noted that the conventions we have referred to are not conventions in the sense that one convention is as good as another. Even the paradigm conventionalists—Quine, Duhem, Poincaré—agree that considerations of simplicity, convenience, familiarity, elegance, and so on can be used to distinguish between one convention and another, and that these grounds are overriding.

Less dramatically and more precisely: laws and theories are features of our scientific language, as are the predicates and relations we take to be observational. To change the set of laws and theories, or the conditions of application of observational predicates, is to change the language. But I shall claim that we can have *epistemic* grounds for such changes, and that these grounds are a direct reflection of inductive considerations in a relatively narrow sense.

Generalizations

Much of the philosophical focus on induction has been concerned with universal generalizations: "All crows are black," "All emeralds are green," "Every creature with kidneys is a creature with a heart," and the like. It has been felt that if we can show how and why (and whether) such generalizations can be given rational justification by our limited observations of the world, we'll be home free. It is felt that it is but a short step from here to quantitative laws (all samples of mercury have a density of 13.6) and an only slightly longer jump to general theories (every mechanical system satisfies Newton's laws). After we have achieved an understanding of general theories, we can go back, in a mopping-up operation, to take care of such unimportant trivia as statistical generalizations.

There may be some truth in this view. Not much, perhaps, for it is hard to be quite sure that there *are* any true universal generalizations of the sort mentioned. Surely not all crows are black; maybe all emeralds are green, and maybe, at the same time, greenness is not one of the defining characteristics of emeralds, but it's hard to be sure. There are creatures with hearts and no kidneys in butcher shops; and the fact that there are none among the functional and untampered with seems more an indirect consequence of higher-order physiological theories than an empirical induction. But no matter. It may yet be instructive to consider a possible case.

We shall draw, in this and subsequent chapters, on the statistical claims made earlier. That is, we shall suppose that relative to a corpus of statements **K**, containing knowledge of a sample of A's of which $r\%$ are B's, it may be highly probable that about $r\%$ of all A's are B's. More generally, suppose that **K** contains knowledge about the distribution of the quantity Q in a subset of the A's. **K** may or may not contain knowledge of the family of distributions to which the distribution of Q in A belongs, and may or may not contain knowledge of the distribution of the distributions of Q among a superset to which A belongs (i.e., **K** may or may not contain the materials for a Bayesian inference concerning the distribution of Q in A). It still may be rendered highly probable, by knowledge of a sample, that the distribution of Q in A is one of a family of distributions F. In general and roughly: the total evidence may render it highly probable that the distribution of Q in A is close to that of Q in our *sample* from A. For present purposes, we must take all this, controversial though it may be, for granted.

What is generally "given" as a basis for an inductive generalization of the form "All A's are B's" is a collection of observations: "This is an A and is a B," "That is an A and is a B," and so on, and perhaps "This is not a B, and is not an A either" and "That is not an A, even though it is a B." Let us take **K*** to be the deductively closed Ur-corpus; suppose that it contains knowledge of a finite set of A's, non-A's, B's, and non-B's, and that 100% of the A's in this set are also B's.

Relative to this **K***, if the sample is random with respect to reflecting the proportion of A's that are B's, it may be highly probable that between $1-\delta$ and 1.0 of of the A's (in general) are B's. That is, "$\%(B,A)\in[1-\delta,1.0]$" may belong to the evidential corpus **K** in virtue of its high probability relative to the incorrigibilia embodied in **K***. This is not yet a universal generalization. But if we suppose that $1-\delta$ is larger than the level of acceptance of the practical corpus **K'**, *and* that there is no statement "$(\exists x)(Ax\ \&\sim Bx)$" in the corpus **K'**, then the net result is the same as that of including the universal generalization "$(x)(Ax\rightarrow Bx)$" in the corpus **K**. We may, in fact, adopt an inductive rule to the effect that when these three conditions are satisfied:

1. "$\%(B,A) \in [1-\delta, 1.0]$" belongs to **K**,
2. "$(\exists x)(Ax \& \sim Bx)$" does *not* belong to **K**,
3. $1-\delta$ is greater than the level of practical certainty characteristic of the corpus **K'**,

then we may include the sentence "$(x)(Ax \rightarrow Bx)$" in **K**. We must, of course, withdraw it as soon as one of the three conditions is violated — for example, when we observe an *A* that is not a *B* or when we come to believe (justifiably and rationally) on other grounds that there is one. Meanwhile, however, the generalization serves (unnecessarily) to warrant the inference from "*Aa*" to "*Ba*."

To visualize these sets of statements, it may help to think of them as displayed in Figure 1. **K***, the Ur-corpus, contains all those sentences of the language **L** that we regard as incorrigible: the truths of logic, the truths of mathematics, whatever principles and generalizations we wish to regard as characteristic of the language **L**, and finally, statements (if any) accepted on the basis of observation.

K, the set of evidential certainties, contains sentences of **L** that are very highly probable relative to **K***, the Ur-corpus. Every sentence in **K*** will also be in **K**, since if *S* belongs to **K***, the probability of *S* relative to **K*** will be [1.0,1.0], which warrants the inclusion of *S* in **K**. But there may be some other sentences as well. What counts as a "high enough" probability to warrant inclusion in the evidential corpus can

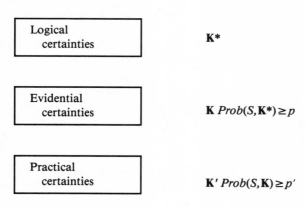

Figure 1

be left to depend on context.[10] We denote that contextually dependent constant by p.

K′ is the set of practical certainties, consisting of those sentences probable enough relative to the corpus of evidential certainties. "Probable enough" is again represented by a contextually determined constant, p'; we impose the condition that $p'<p$. As before, **K** will be a subset of **K′**.

Let us now confront the uncertainties of reality. Generalizations based on enumerative evidence, if they exist at all, are not often based on the observations of a single person. It is not that I have examined thousands of crows or emeralds, or dissected thousands of creatures, but that I have examined some, my friends have examined some, their friends have examined some, and so on. Among all these reports, is there not *one* that testifies to the existence of a counterexample to the generalization in question? If I myself, personally, have accumulated all the data, have I never been moved to write in my notebook, "On the sixth inst. there did appear, though but briefly, on the ridge of the southwest barn, a crow of most unusual color?"

Taking all of these reports and observations at face value, there must appear in **K** (and therefore in **K′**) some instantial statement of the form "$Aa\&\sim Ba$"; and by limited deductive closure, there will thus appear both in **K** and in **K′** the sentence "$(\exists x)(Ax\&\sim Bx)$," precluding, by our earlier proposal, the inclusion of the universal generalization "$(x)(Ax\rightarrow Bx)$" in **K**. Note that we need not assume that the counterinstantial observation is veridical. It need merely have enough intrinsic likelihood to warrant its inclusion in the corpus **K** on which our inference is to be based.

Uncertainty and error in the observational foundation for a universal generalization appear to undermine the rationale for incorporating universal generalizations in a corpus **K** of evidential certainties. If only we had some way of casting special doubt on those putative counterexamples that stand between us and the acceptance of a nice generalization!

Well, we have one good ground for not taking them seriously: they interfere with our acceptance of a universal generalization. Our strategy can be to regard the alleged observations of counterexamples as erroneous. The reconstruction in this case consists of two parts: a shift

---

[10]For determining "level" by context, see "Full Belief," *Theory and Decision* **25**, 1988, 137-162.

to a new language, of which "$(x)(Ax \to Bx)$" as an a priori feature is one part. The other consists in the acceptance, in a corpus of the corresponding metalanguage, of a statistical statement reflecting the frequency with which the relevant observational judgments must be taken to be in error. Both parts deserve comment.

To accept "$(x)(Ax \to Bx)$" as an a priori feature of the language is to go beyond even the claim that 100% of the A's are B's: it is to claim not only that any A is in fact a B, but that any A *must* be a B. By this move we achieve not only universal generality but nomic necessity, or at least a linguistic counterpart thereof.[11]

The other aspect of the move is more complex. It involves the same ideas that were introduced in connection with errors of measurement, but these ideas are worth reviewing in this new context. We have already added to our metalanguage the predicate "O" (for "observed"), and in order to avoid having to consider the reliability of memory, we interpreted "$O(X,S,t)$" to mean that X (person, society, or whatever) has been moved to accept — has judged true, has written in a laboratory notebook — the statement S of our object language at some time before t, and this on observational grounds. We continue that discussion, started in Chapter 5, here.

We write

$$O(X,S) = \{S : (\exists t)(O(X,S,t))\}$$

for the set of statements X is moved to accept at some time or other. Since we also include the object language as a (proper) part of the metalanguage, and so can make the distinction between veridical and erroneous observational acceptance, we may write

$$VO(X,S) = \{S : S \in O(X,S) \& S\}$$

for the set of X's veridical observations and

$$EO(X,S) = \{S : S \in O(X,S) \& \sim S\}$$

for the set of X's erroneous observations.

For referring to particular kinds of basic sentences (a "basic sentence" is an atomic sentence or its negation), let us write $O^A(X,t)$ for the set of sentences of the form ⌜Ay⌝ accepted by X on the basis of observation before t, $O^{nB}(X)$ for the set of sentences of the form ⌜$\sim Bw$⌝ accepted by X at any time on the basis of observation, and so on.

[11] Bas van Fraassen, "The Only Necessity Is Verbal Necessity," *Journal of Philosophy* 74, 1977, 71-85.

If X's language includes "$(x)(Ax \to Bx)$" and $O(X,"Aa",t)$ and $O(X,"\sim Ba",t)$—that is, X has accepted both "Aa" and "$\sim Ba$" before time t on the basis of observation—we know (and he knows) that at least one statement X is moved by observation to accept is false: either "Aa" or "$\sim Ba$." But note that we may not know which. The generalization is not open to question, since it is merely a feature of X's language.

Consider the set of sentences X is moved to accept at t by observation:

$$O(X,t) = \{S : O(X,S,t)\}.$$

The a priori constraints of the language of X's corpus may require that some be in error, but need not specify which, as just illustrated. How many? We could, of course, regard observation as totally unreliable, since in X's language it admits of error. But that's throwing away the baby.

Instead, let us adopt the principle that observations are not to be impugned without reason. This principle has a lot of intuitive appeal, but it is not easy to provide an argument in its favor. It does seem to beg the question against skepticism, but it is not clear that in the absence of such a principle, language would be at all possible. The principle, succinctly put, is, "Don't ask for trouble!" More explicitly:

> Given a set of observational judgments, assume that they contain no more error than is required to render them consistent with the a priori principles of your language.

This is the *minimization principle*, introduced earlier.

With the help of this principle, we can compute the minimum frequency with which statements in $O(X,t)$ are erroneous, that is, belong to $EO(X,t)$. Since this gives us the relative frequency of error at t, we may, provided that the right conditions of randomness are met, use it to infer, with high probability, the general frequency of observational error:

$$\%(EO(X), O(X)) \in [p,q]$$

In general, we may reasonably suppose that the conditions of randomness *are* met and that we *can* make this statistical inference.

But this isn't the inference we most want. In the case at hand, it is only statements of the forms $\ulcorner Ax \urcorner$ and $\ulcorner \sim Bx \urcorner$ that need to be regarded as subject to error. We would like the error frequencies of these sorts of statements. In the example, the statements come in pairs, and we can obtain consistency by regarding *either* member of the pair as erroneous. To get specific error rates for statements of the two forms that concern us, we must invoke another principle:

> Given the satisfaction of the minimization principle, attribute error to the various kinds of basic statements in such a way as to distribute error as equally as possible among them.

We referred to this earlier as the *distribution principle*. It may also be thought of as a kind of second-order minimization principle in which we seek to minimize the maximum error rate.

For example, suppose that we have 100 instances of $O^A(X,t)$ and 100 instances of $O^{nB}(X,t)$, and that there are 10 terms z such that both $O(X, \ulcorner Az \urcorner, t)$ and $O(X, \ulcorner \sim Bz \urcorner, t)$. We must reject as erroneous at least one member of each pair. So, by the distribution principle, we should regard five of each as in error, yielding a *sample* frequency of error of 5% for each kind of statement, rather than a 10% error for one kind and a 0% error for the other. Note that we do not have to settle on any *particular* statements of the 10 troublesome pairs to be regarded as false. From these data we can infer that *about* 5% of each kind of observation are erroneous.

Let us now look at the two cases: the language with the generalization and the language without the generalization. Let L_1 be our original language, in which we suppose our observations incorrigible, and let L_2 be the alternative language in which we accept as an a priori truth "$(x)(Ax \rightarrow Bx)$." K_1^* and K_2^* are the incorrigible (and deductively closed) corpora in the two cases, MK_1^* and MK_2^* the incorrigible metacorpora, K_1 and K_2 the evidential corpora (at level p), MK_1 and MK_2 the metacorpora of level p, and K_1' and K_2' the practical corpora of level $p' < p$. Figure 2 gives the general picture for a single language, L_1. Figure 3 gives the corresponding picture for the language L_2 embodying the a priori generalization that all A's are B's.

In the case of L_1, since we assume that all observations are veridical, we have a lot of statements of the form $\ulcorner Ay \urcorner$, $\ulcorner Bw \urcorner$, $\ulcorner \sim Az \urcorner$, and $\ulcorner \sim Bx \urcorner$ in K_1^*. We must also admit that for a few terms w, we have both $\ulcorner Aw \urcorner$ and $\ulcorner \sim Bw \urcorner$ in K_1^*. K_1^* also contains mathematical and logical theorems, and the like.

In the case of K_2^* we have only logic and mathematics; no observations are regarded as incorrigible. MK_1^* contains for each nonlogical basic sentence S in K_1^*, $O(X,S,t)$ and MK_2^* contains exactly the same sentences. MK_2^*, but not MK_1^*, contains (in virtue of deductive clo-

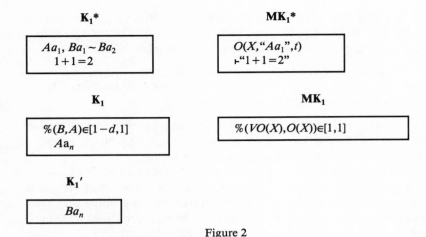

Figure 2

sure and our two principles) the assertion that of the statements $O^A(X,t)$, such and such a percentage belong to $EO^A(X,t)$, and that of the statements $O^{nB}(X,t)$, such and such a percentage belong to $EO^{nB}(X,t)$. In MK_1^*, in virtue of our minimization principle, we have the corresponding assertions that none of the statements of these forms are in error.

Proceeding to the next level, we may use the sample frequencies in MK_2^* as a basis for inferring in MK_2 the statistical statements:

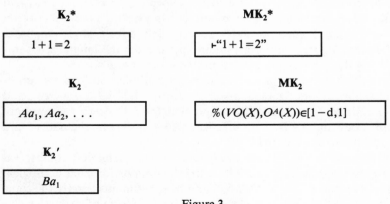

Figure 3

$$\%(EO^A(X), O^A(X)) \in [p^A, q^A]$$
$$\%(EO^{nB}(X), O^{nB}(X)) \in [p^{nB}, q^{nB}]$$

and so also

$$\%(VO^A(X), O^A(X)) \in [1-q^A, 1-p^A]$$
$$\%(VO^{nB}(X), O^{nB}(X)) \in [1-q^{nB}, 1-p^{nB}].$$

For this example, we suppose that $q^A = q^{nB} < 1-p$, where p is the level of the evidential corpus.

There are special subclasses of $O^A(X)$ and $O^{nB}(X)$ that will also concern us: the subclass of $O^A(X)$ of statements of the form ⌜Aw⌝ that are paired with statements of the form ⌜$\sim Bw$⌝ for the same term w, and the subclass of $O^{nB}(X)$ of statements of the form ⌜$\sim By$⌝ that are paired with statements of the form ⌜Ay⌝ for the same term y. Denote these subclasses by CO^A and CO^{nB}, respectively. Our distribution and minimization principles ensure that we will regard about half of such statements as false, so that we may infer in $\mathbf{MK_2}$:

$$\%(VO^A, CO^A) \in [1/2 - c_1, 1/2 + d_1]$$
$$\%(VO^{nB}, CO^{nB}) \in [1/2 - c_1, 1/2 + d_1]$$

For similar reasons, back in the object language $\mathbf{L_1}$, the statistical statement

$$\%(B, A) \in [1-d, 1]$$

belongs to $\mathbf{K_1}$, where $1-d > p'$, since frequency data about A's and B's are provided in $\mathbf{K_1^*}$. $\mathbf{K_1}$ also contains all the observation statements.

$\mathbf{K_2}$ remains to be considered at this level. Suppose that $O(X, ⌜Aw⌝, t)$ belongs to $\mathbf{MK_2^*}$. Then if the sentence ⌜Aw⌝ is a *random member* of $O^A(X)$ with respect to belonging to $VO^A(X)$, the probability of ⌜Aw⌝$\in VO^A(X)$ is $[1-q^A, 1-p^A]$ relative to $\mathbf{MK_2}$, and (in view of the presence of $O(X, ⌜Aw⌝, t)$) that is also the probability of ⌜Aw⌝ relative to $\mathbf{MK_2}$. But this warrants the inclusion of ⌜Aw⌝ in $\mathbf{K_2}$. Similar remarks hold for ⌜$\sim By$⌝ when it is a random member of $O^{nB}(X)$ with respect to $VO^{nB}(X)$.

But randomness does not always hold. Suppose that for some term w, both $O(X, ⌜Aw⌝, t)$ and $O(X, ⌜\sim Bw⌝, t)$. Then ⌜Aw⌝ is known to belong to a subset of $O^A(X)$ in which the minimum relative frequency of statements belonging to $VO^A(X)$ is less than one-half, namely, CO^A.

Thus, in the case of *pairs* of sentences of the forms ⌜Aw⌝ and ⌜$\sim Bw$⌝, neither will be probable enough to be included in $\mathbf{K_2}$.

On the other hand, if ⌜Aw⌝ is included in $\mathbf{K}_2$, so, in virtue of the generalization characteristic of $\mathbf{L}_2$, will be ⌜Bw⌝; and similarly, if ⌜$\sim Bw$⌝ is included in $\mathbf{K}_2$, so also will be ⌜$\sim Aw$⌝.

Finally, we should look at the corpora of practical certainties. $\mathbf{K}_1'$ will contain all the statements of $\mathbf{K}_1$, together with statements whose probability relative to $\mathbf{K}_1$ is greater than p', the acceptance level of the practical corpus. Similarly, $\mathbf{K}_2'$ will contain all the statements of $\mathbf{K}_2$ together with those that are highly probable relative to $\mathbf{K}_2$.

If w is a random member of A with respect to B, relative to $\mathbf{K}_1$, then the probability of ⌜Bw⌝ will be $[1-d,1.0]$, and since $1-d>p'$, ⌜Bw⌝ will belong to $\mathbf{K}_1'$. And w will be a random member of A in the required sense, unless we already know ⌜Bw⌝ or unless we know ⌜$\sim Bw$⌝. But since "$\%(B,A)\in[1-d,1,0]$" does not imply anything about "$\%(\sim A,\sim B)$," $\mathbf{K}_1'$ will not, on the basis of anything we have said, contain ⌜$\sim Aw$⌝ when $\mathbf{K}_1$ contains ⌜$\sim Bw$⌝.

On the other hand, if $\mathbf{K}_2$ contains ⌜$\sim Bw$⌝, it will automatically contain ⌜$\sim Aw$⌝, and therefore so will $\mathbf{K}_2'$. But remember that $\mathbf{K}_2$ contains fewer statements of the forms ⌜Aw⌝ and ⌜$\sim By$⌝ than does $\mathbf{K}_1$.

It is worth noting that in this very simple case there is a generic difference between the corpora generated under $\mathbf{L}_1$ and the corpora generated under $\mathbf{L}_2$. Under $\mathbf{L}_1$, the metacorpora $\mathbf{MK}_1^*$ and $\mathbf{MK}_1$ serve no useful function; they aren't empty, but they do not influence the contents of $\mathbf{K}_1$ and $\mathbf{K}_1'$. Thus, if we were to speak a language containing only incorrigible observation sentences, the picture presented in Figure 4 would capture all the relevant detail.

On the other hand, in $\mathbf{L}_2$, where all of our observation sentences are regarded as corrigible, that is, subject to error and uncertainty, the closed observation corpus $\mathbf{K}_2^*$ serves no useful function, containing only the a priori truths of mathematics and the language $\mathbf{L}_2$. We would picture our corpora as in Figure 5 on p. 126.

To get a better feel for what is going on, and to develop a principle for choosing between $\mathbf{L}_1$ and $\mathbf{L}_2$, we shall look at some hypothetical numbers in the next section.

At this point, it suffices to emphasize that what is driving the choice between conventions is the notion of error. We need never (in principle) admit to error; our bodies of knowledge would be empty of useful content. We are also free to imagine whatever science fiction world we like; the consequence would be that our observational experience would have to be regarded as embodying intolerable amounts of

$$\boxed{\begin{array}{l} Aa_1, Ba_2, \sim Ba_k, \ldots \\ 1+1=2, Aa_n \end{array}} \qquad K_1^*$$

$$\boxed{\begin{array}{l} \%(B,A) \in [1-d, 1] \\ Aa_1, Ba_2, \sim Ba_k, \ldots \\ 1+1=2, Aa_n \end{array}} \qquad K_1$$

$$\boxed{\begin{array}{l} \%(B,A) \in [1-d, 1] \\ Aa_1, Ba_2, \sim Ba_k, \ldots \\ 1+1=2, Aa_n \\ Ba_n \end{array}} \qquad K_1'$$

Figure 4

error. Reason and science (not to mention practicality!) dictate a middle course.

Since the notion of error is fundamental, the two principles that guide us in our choice of error functions — the minimization principle and the distribution principle — assume a deep philosophical role. They are not uncontroversial; in particular, one may prefer to seek *explanations* for errors rather than *data about* errors.[12]

In general, of course, we have a solid framework of knowledge about the physical world; we have a great deal of statistical information about the frequencies, magnitudes, and sources of observational error; and that is what we use in assessing the likelihood of observational error in actuality. It is clear that one would learn quickly, even in the kind of qualitative case being discussed, that observations made under some circumstances are "better" than those made under other circumstances. For example, we may consider the A-observations made under good light as a distinct subset of A-observations, and we may find that the rate of error we are obliged to impute to these observations is less than that required in general. We have here a classical statistical problem of deciding when it is useful to distinguish

[12] This idea has been proposed by Paul Scatena, *APA Western Division Meeting*, 1988, in "Principles of the Theory of Error."

$$\boxed{\begin{array}{l} O(X,\text{``}Aa,\text{''},t) \\ O(X,\text{``}Aa,\text{''}t) \\ O(X,\text{``}\sim Ba,\text{''}t) \\ \vdash \text{``}1+1=2\text{''} \vdash \text{``}(x)(Ax \to Bx)\text{''} \end{array}} \quad \text{MK}_2\text{*}$$

$$\boxed{\begin{array}{l} \%(B,A)\in[1-d,1] \\ Aa_1, Ba_2, \sim Ba_k, \ldots \\ 1+1=2, Aa_n \end{array}} \quad \text{K}_2 \qquad \boxed{\begin{array}{l} \%(VO(X),O(X))\in[1-e,1] \\ Aa_n \sim Ba_k \end{array}} \quad \text{MK}_2$$

$$\boxed{\begin{array}{l} \%(B,A)\in[1-d,1] \\ Aa_1, Ba_2, \sim Ba_k, \ldots \\ 1+1=2, Aa_n \\ Ba_n \end{array}} \quad \text{K}_2'$$

Figure 5

between two populations. The two principles have substantive import only when we are dealing with very primitive and fundamental questions. They illuminate the theory of observational error, and they provide a starting point for that theory, but they rapidly become overlaid with substantive statistical knowledge about error. We shall explore some of the ways in which this happens in later sections of this chapter.[13]

Choosing between Two Languages

In discussing our baby example, I have been supposing that there is some attractiveness to replacing the language that lacks the generalization that all A's are B's by the language that incorporates that generalization. Clearly, that is not always the case. Suppose, in fact, that our experience has consisted of observations of each of the four categories

[13]More details will be found in *Theory and Measurement*, Cambridge University Press, Cambridge, 1984.

of things in roughly equal numbers. The move to the new language would then involve assuming that a lot of our observations were in error, and there would be no compensating gain. In fact, it could entail such high predicted error frequencies that some kinds of observations would never be probable enough to be accepted in any rational corpus of reasonably high level.

To make all this clearer, and to formalize the epistemic grounds for preferring one language to another, let us consider the frequencies with which various basic statements appear in the various corpora. Suppose that the number of pairs of basic statements of the forms ⌜Bw⌝ and ⌜Aw⌝ for a common term w is f_1 and so on, as in the following list. The fifth through eighth entries give the frequencies of basic statements that do not occur as one of a pair: for example, f_5 is the cardinality of the set of statements of the form ⌜Aw⌝ that occur in $\mathbf{K}_1$ unaccompanied by either ⌜Bw⌝ or ⌜$\sim Bw$⌝.

⌜Aw⌝	⌜Bw⌝	f_1
⌜Aw⌝	⌜$\sim Bw$⌝	f_2
⌜$\sim Aw$⌝	⌜Bw⌝	f_3
⌜$\sim Aw$⌝	⌜$\sim Bw$⌝	f_4
⌜Aw⌝		f_5
⌜$\sim Aw$⌝		f_6
⌜Bw⌝		f_7
⌜$\sim Bw$⌝		f_8

In our example we assumed, rather vaguely, that f_1 was large and that f_2 was small. Here we must be more precise. Let us fix p and p', the level of evidential certainty and the level of practical certainty, as before. Given a sample of N Z's of which M are W's, we may infer with probability p (assuming that the sample is random in the epistemic sense, which is no problem in this simple case) that the proportion of Z's that are W's in general lies between $M/N-j(N)$ and $M/N+k(N)$. The exact functions j and k do not concern us, though it may help our intuitions to know that they are both of the order of $N^{-1/2}$. Below we shall even suppress their arguments.

Consider first the case of completely known instances—that is, where f_5 through f_8 are zero. Assume that the only substantive statistical statement warranted at the p-level is "%$(B,A)\in[f_1/(f_1+f_2)-j, f_1/(f_1+f_2)+k]$" and that $f_1/(f_1+f_2)-j>p'$. Then:

K_1^* contains $f=f_1+f_2+f_3+f_4$ empirical statements — our observations.

MK_1^* contains f metalinguistic statements $O(X,S,t)$, where S is one of the observation statements in K_1^*, and, embodying the fact that L_1 does not allow for observational error, the statement "$(S)(S\in O(X,S) \to S \in VO(X,S))$."

K_1 contains the same evidence statements as K_1^*, together with the statistical statement just mentioned.

MK_1 contains the same statements as MK_1^*.

K_1' contains the same evidence statements as K_1^* and a slightly stronger statistical statement (since $p<p'$).

In the corpora corresponding to the second language, L_2, we have:

K_2^* contains only logical and mathematical truths — including "$(x)(Ax \to Bx)$" — but no observations.

MK_2^*, like MK_1^*, contains f metalinguistic statements $O(X,S,t)$, where S is one of our observation statements. It contains no statement about error.

MK_2 contains our knowledge of error:

"$\%(VO^A(X), O^A(X)) \in [(f_1 + \frac{1}{2}f_2)/(f_1+f_2) - j, (f_1+\frac{1}{2}f_2)/(f_1+f_2)+k]$"
"$\%(VO^{nB}(X), O^{nB}(X)) \in [(f_4 + \frac{1}{2}f_2)/(f_4+f_2) - j, (f_4+\frac{1}{2}f_2)/(f_4+f_2)+k]$"

since by the distribution principle only half of the f_2 observations in the second category are impugned by the generalization. Writing $O^{AnB}(X)$ for the set of A-observations accompanied by a $\sim B$-observation of the same object, we also have,

"$\%(VO^{AnB}(X), O^{AnB}(X)) \in [\frac{1}{2}-j, \frac{1}{2}+k]$"

"$\%(VO^{nBA}(X), O^{nBA}(X)) \in [\frac{1}{2}-j, \frac{1}{2}+k]$"

K_2 contains the AB, $\sim AB$, $\sim A \sim B$ observations, since none of these can be impugned. But all of the $A \& \sim B$ statements are suspect; and in fact, they have lower probabilities of less than $\frac{1}{2}$! We suffer a loss of f_2 statements compared to K_1. The same loss is transmitted to K_2'.

In the case of completely known instances, there is no advantage to L_2 as opposed to L_1. How can we explain, then, our lust for generality? Consider limited knowledge.

Suppose that $f_1, \ldots f_8$ are all positive; then:

K_1^* will contain all the observation statements.

MK_1^* will contain the observation reports.

K_1 will contain the observation statements from K_1^*, together with the statistical statements previously described. Note that the bare observation of A's, unaccompanied by observations of whether they are B's or not, does not contribute to the statistical data on which we infer the frequency of B's among A's.

MK_1 contains the same observation reports as MK_1^*.

K_1' now contains novelty. It contains the f observation statements of K_1^* and, in addition, f_5 new statements of the form $\ulcorner Bw \urcorner$. This is because everything of which it is known only that it belongs to A will be, relative to K_1, a random member of A with respect to being in B, and the probability that it is a B will be high enough (in accord with the statistical statement in K_1) to warrant the inclusion of the statement that it is a B in K_1'. K_1' therefore contains $f+f_5$ basic statements.

K_2^* contains nothing interesting.

MK_2^* contains observation reports, as before.

MK_2 will contain these reports, together with statistical statements reflecting the reliability of those reports. Note again that partial observations will not contribute to the statistics.

K_2 will be impoverished, as before, by the f_2 statements that are known to contain error.[14] But now that we have a generalization in our language, K_2 gains f_5 statements of the form $\ulcorner Bw \urcorner$. (These are the statements that appeared by statistical inference in K_1'.) In addition, however, K_2 gains f_8 statements of the form $\ulcorner \sim Aw \urcorner$. Note also that these statements count as *evidential* certainties.

The basic fact underlying this phenomenon is that universal generalizations are convertible (from "All A are B" we can obtain "All $\sim B$ are $\sim A$"), but statistical generalizations (even "nearly universal" ones) are not. "Almost all A are B" is perfectly compatible with "Almost all $\sim B$ are A."

A similar gain in power comes from considering several generalizations. If we know that all A's are B's and that all B's are C's, we can infer that a particular A is a C. But if our knowledge is merely statistical, no such inference follows, even in probabilistic form, since it does

---

[14]There *may* be further loss. If $(f_4+\frac{1}{2}f_2)/(f_4+f_2)-j$ is not greater than p, then we do not have grounds for taking nB observations to be evidentially reliable; a random nB observation does not have a high enough probability of being veridical to be included in K_2. (A random $nB\&nA$ observation, however, is acceptable; we know that we will not need to reject that observation. Using the notation introduced earlier, it is a random member of $O^{nBnA}(X)$, of which none have needed to be rejected.) Note how un-Bayesian this is: a conjunction is more reliable than one conjunct.

not follow from "Almost all A's are B's" and "Almost all B's are C's" that *any* of the A's are C's.

What really interests us in our bodies of knowledge is their *predictive content*. That is, statements (in our example) of the first four categories—completely known instances—are water over the dam. They represent history (in the dull sense) rather than prediction or control of the future. It is the predictive content of our corpus of practical certainties that constitutes the epistemological advantage, if any, of moving from L_1 to L_2.

To regularize the measurement of this predictive content, let us consider a corpus containing no partial knowledge. That is, let us purge what partial knowledge there may be from **MK***. Then let us add partial knowledge in the form of basic sentences concerning a finite hypothetical sequence of individuals, and let us do so with frequencies reflecting the marginal frequencies among the fully known instances. Thus, in the example, one would add reports of the forms $\ulcorner Aw \urcorner$, $\ulcorner Bw \urcorner$, $\ulcorner \sim Aw \urcorner$, and $\ulcorner \sim Bw \urcorner$ in the ratios $f_1+f_2 : f_1+f_3 : f_1+f_4 : f_2+f_4$.

We now offer the following conventional choice principle:

> Language L_2 is *epistemically preferable* to L_1 just in case when we add enough incomplete observation reports (basic sentences) to both MK_1^* and MK_2^*, in the historical ratio, the corpus of practical certainties K_2' contains more basic observation sentences than the practical corpus K_1'.

By "enough" we intend that there be an N such that if we add more than a total of N, the condition is always satisfied. By "incomplete" we intend instances about which there is something left to observe, as in the last four categories of our example. Since the levels of evidential certainty, p, and practical certainty, p', are parameters in this characterization of epistemic preference, they should strictly form part of the definiendum. Similarly, we have no guarantee that the relation between the two languages will not reverse itself infinitely often as we increase the total number of incomplete observations. But one way to handle these problems—one I take to be adequate in the present context—is simply to say that when we have reversals for levels of practical or evidential certainty of interest, or for large numbers of partial observations, there simply is no epistemic preference relation between the two languages.

At a given level of practical certainty, a language that admits of error in its application but that incorporates universal generalizations can, with the right input, lead to a more populous collection of predictive basic observation statements. This yields an epistemic criterion that serves to judge between conventional languages.

Refutation

One classical and well-known view of the role of universal generalizations in science requires that they be refutable in order to be credited with empirical content. Popper, for example, has made this the cornerstone of his philosophy of science.[15] Another view, the hypothetico-deductive view, requires that refutable empirical statements be deducible from hypotheses under test. Having given up "verification" as applied to scientific hypotheses, the descendents of the old-time positivists have required that scientific generalizations, to be respectable, must be refutable by some possible set of observations. On the view that I have been describing, this requirement is not met. Once we have included the generalization "$(x)(Ax \rightarrow Bx)$" as an a priori feature of our language, no possible observations can refute it. Such observations reflect only on the relevant frequency of error. We may "observe" an A that is not a B—that is, we may include "O(us, "$Aa\& \sim Ba$",t)" in our metacorpus **MK***—but that in itself will not lead us to reject the generalization. It will merely contribute to the statistical evidence on the basis of which we determine the reliability with which we can apply the predicates "A" and "B" and their complements.

We have seen that if the observational reliability falls low enough, the language that does *not* incorporate the feature that all A's are B's will be preferable. In a richer context, this can be made to look a lot like testing and refuting a universal generalization.

This comes about through the operation of the epistemological notion of randomness. Suppose that we have a language with a number of generalizations involving A's, B's, $\sim A$'s, $\sim B$'s, and a large number of other predicates, but that does not contain the generalization that all A's are B's. By enumerative induction, using the fact that the observational application of "A" and "B" is highly reliable, we obtain acceptance in **K** of "$\%(B,A) \in [1-r, 1.0]$."

Under suitable circumstances, this warrants considering the move to a language in which "$(x)(Ax \rightarrow Bx)$" is an a priori feature. But,

---

[15]K. Popper, The *Logic of Scientific Discovery*, Hutchinson, London, 1959. (First German edition, 1934.)

perhaps influenced by Popper, we feel obliged to explore more thoroughly the consequences of using this language. We seek out a number of A's and $\sim B$'s. And it turns out that there are circumstances C under which we are quite often moved both to accept ⌜Aw⌝ and to deny ⌜Bw⌝. Since A and B are involved in many generalizations, this need not have much of an effect on the *general* reliability of A- and B-judgments.

Now the probability that an A-judgment is veridical depends on the reference class to which that judgment is referred and on our knowledge of the frequency of veridical judgments in that reference class. What we have discovered is that *in the circumstances C we have sought out*, the frequency of errors in A-judgments and in $\sim B$-judgments must be regarded as large. Thus, in these circumstances, we cannot accept "Aa" and "$\sim Ba$" on the basis of observation if we use the language containing the generalization in question. In the case envisaged, our practical corpus is impoverished rather than enriched, so far as these special circumstances are concerned, by the inclusion of the generalization that all A's are B's.

But there are two considerations that can change this situation. Of course, if we can specify the circumstances C in our language, and tell *reliably* when they obtain, then we can usefully accept "$(x)(\sim Cx \rightarrow (Ax \rightarrow Bx))$" into our language and have the best of both possible worlds. More interestingly, it is possible that the generalization "$(x)(Ax \rightarrow Bx)$" plays a role in a web of generalizations that as a whole greatly enriches our practical corpus. That is, it might be that if we accept the impoverishment caused by the fact that we cannot reliably tell A's and $\sim B$'s by observation in the special circumstances that appear to "refute" the generalization, there is a powerful compensating gain.

It would clearly be helpful and enlightening to be able to give a worked-out set of realistic examples in which these various outcomes can be exhibited. The example of white and black swans, discussed in "All Generalizations are Analytic,"[16] illustrates the gambit of restricting generalizations: when we become convinced that there are swans in Australia that people very frequently judge to be black, we must suppose either that judgments of blackness and swanhood are prone to a

---

[16]H. Kyburg, *Epistemology and Inference*, University of Minnesota Press, Minneapolis, 1983, essay 19.

lot of error in Australia or that it is more constructive to move to a language characterized by "All but Australian swans are white" and "All Australian swans are black," rather than simply "All swans are white." Note that whether or not this is constructive is a matter of ornithology, not perceptual psychology.

There are also cases in which, with the help of rich theoretical background knowledge, generalizations can be quickly refuted. At one time, it was reasonable to believe that all (normal) angus cattle were black. This generalization, in view of the qualification "normal," was acceptable as a feature of our language. But efforts were made to breed red angus, and with our knowledge of genetics it wasn't hard to know when they had succeeded. At that point, we had to back off from the "all angus are black" language. It was not the birth of the first red-colored angus, though, that refuted the generalization. (The generalization that all cows have three stomachs is not refuted by an occasional genetic anomaly.) It was the fact that there was a nontrivial number of red angus, and particularly the fact that careful breeding could produce more.

The more complex case, in which the generalization plays a major role in a web of theory, is harder to illustrate. As a somewhat far-fetched example, suppose that it becomes acceptable that every strain in a certain species has an observable property A. Furthermore, it is acceptable that everything that has the observable property A also has the observable property B, and that whatever has B also has C. (All fleas jump, and everything that jumps has flat feet; and everything that has flat feet spends a lot of time just sitting.)

Let the generalization that all A's are B's be brought into question by the recording of (alleged) instances of A's that are not B's. Yet it may also be the case that the observable property C is present in each strain of the species concerned. (All fleas spend a lot of time just sitting.) Now it may be that in this sort of case, our general knowledge of genetics makes it preferable to move to a language in which $\sim B$ is no longer as observable as it was. Rather than saying that we should reject the generalization that all A's are B's, and with it some significant fragment of genetic theory, we say that in some of these species B is not easily observable or even not observable at all. (Some species of fleas have a foot structure that makes it difficult to tell whether or not they are flat-footed. But since they all like to sit around a lot, and since we have evidence to the effect that flat-footedness is related to the same genes that cause sitting around, we suppose that it is just hard to tell about their feet.)

It is hard to find rich theories in which universal generalizations involving observation predicates play a significant role. This is a reflection of an approach to scientific generalization already apparent in our example: an approach that replaces incorrigible observations by observations that are only statistically reliable. Real science—what we are really concerned with—is essentially quantitative in character. Although many examples of exciting modern science that we find in the journals appear to be qualitative ("We have just verified the existence of the such-and-such particle!"), these claims rest heavily on *quantitative* observation. It is in the realm of quantitative laws and theories that the structure proposed pays off most significantly.

8

Laws and Theories

Quantities and Errors of Measurement

Central to laws and theories are quantities, measurement, and error. Although those notions have been dealt with in earlier chapters, a brief recapitulation is appropriate here.

"Quantities" in general are to be construed in the same way as "random quantities" or "random variables" are construed in statistics: namely, as functions from a domain to a special set of objects, often taken as a subset of the real numbers. What makes the range of these functions special is the fact that it reflects to some extent — whether or not it is actually real numbers or some physical quantity — the structure of a set of mathematical objects. Thus the quantity *"mass of"* has as its domain ponderable objects, and as its range masses, construed as sets of equiponderable objects. The set of masses has a structure like that of the nonnegative reals, so that masses can be denoted by terms of the form $r*M$, where M represents a selected *"unit"* mass; masses can be added and multiplied by nonnegative reals, as well as by other quantities.[1]

Measurement, correspondingly, is not a process of "assigning numbers to objects," but rather of determining the values of quantity functions applied to objects: more briefly, of assigning magnitudes to objects. This process is always subject to error. Furthermore, this error varies from measurement to measurement and is characterized by a certain distribution. The distribution of error characteristic of a cer-

---

[1]Strictly, the "mass of" function does *not* reflect the structure of the reals. Quantum mechanics suggests that there is only a finite number of distinct masses smaller than a given mass, rather than the uncountable infinity suggested by real analysis.

tain class of measurements is, like all statistical distributions, unknown in a strict sense, but, again like many other statistical distributions, can be known approximately. That is, it can become highly probable that the distribution of errors is one of a family of distributions that we can specify in a useful way.

Typically, the distribution is one according to which the mean of the error is close to zero—positive and negative errors roughly cancel each other out—and according to which large errors are less frequent than small errors. As a useful approximation—and remember that it is an approximation—we often suppose that errors are distributed normally with a mean of 0.0 and a variance that is characteristic of the technique of measurement we are considering.

Of course, we must not take this approximation too seriously. According to any normal distribution, given any real number, there is a finite probability that the error will be greater than that number and a finite probability that it will be less than (more negative than) that number. Thus, if I weigh a ball bearing and obtain 1 gram as the result, there is a finite probability that its true weight is 10 kilograms, and a finite probability that its true weight is negative 10 kilograms, according to a normal distribution. This is no problem if we remember that the normal distribution is just a handy way of representing the family of distributions to which we know the true distribution of errors belongs.

How do we get to know these error distributions? The standard procedure is by calibrating our measuring instruments against known magnitudes. But there are few known magnitudes (for many years, the meter stick at Sèvres was regarded as "known," but now most of the known magnitudes are physical constants, and that they *are* constants presupposes a fair amount of physical theory). From the present point of view, the fact that we confirm theories generally by measurement, and that the units of measurement depend for their constancy on our theories, is not disastrously circular, but quite all right, though that is hard to explain from a refutational point of view.

In fact, from almost any ordinary point of view, the basic source of the error distribution to be associated with a particular method of measurement is a mystery. In real life, we calibrate against certified standards. But except in the context of a theory that warrants the adoption of those standards, it is difficult to understand how standards can be certified. That understanding is one thing that the present philosophy of science is supposed to provide.

It has been argued at length elsewhere[2] that we can get the show on the road in a noncircular way with the help of the same principles alluded to earlier: the conservative principle that we should not attribute more errors to our observations than we need to, and the distribution principle, which says that we should, subject to the conservation principle, play no (unjustified) favorites with regard to the sources of error. What determines the amount of error we *must* introduce? The structure of the language. Let us consider once more an elementary example from measurement theory.

Quantitative Laws

Some writers seem to suggest that the transitivity and additivity of length are just analytic consequences of the very concept of length. If there are concepts (I am quite unsure what they are), and if they are so precise as to yield "analytic" consequences, it seems that they must be learned; and one learns the concept of length by measuring and by reconciling the results of one's measurements with such principles as transitivity and additivity and a number of physical laws. Thus it seems that referring to the *concept* of length adds nothing to what we can learn from the way that length magnitudes are used and the way in which they are approximated in the measuring process.

It is hard for us to think of the additivity law as empirical, though it is clearly quantitative. If we measure x, measure y, and measure their collinear juxtaposition and fail (as we almost always will) to obtain results that satisfy the law, we are inclined to attribute the failure to errors of measurement. If this is not plausible—if the differences are too large, that is, if it is too improbable that we should have made errors so large—then we may question the alleged collinearity or the alleged juxtaposition. Even if in some class of cases we find the law systematically violated, we will wriggle out of refutation. We will say that the bodies concerned changed length in a systematic way.

But this may be regarded as a reflection of the fact that our bodies of knowledge contain so much information that points to the preferability of a language embodying this law that no alternative language is

---

[2] A more detailed analysis of the theory of measurement and its relation to scientific theories in general will be found in my *Theory and Measurement*, Cambridge University Press, Cambridge, 1984.

intuitively feasible. We may still ask what this information is and how it "points."

Length is additive, right? So if we take x to be 4 meters long and y to be 6 meters long, we should take the collinear juxtaposition of x and y to be 10 meters long. But obviously, if we measure x and y and their juxtaposition, we will not get a set of observations satisfying this relation. Explanation: error. But we introduce the minimum amount of error necessary to reconcile our observations with the additivity of length, and furthermore, we assign errors to all three observations. This amounts to taking the additivity of length to be a feature of our language.

Conventional ways of measuring length often presuppose this very law. A meter stick, after all, is the collinear juxtaposition of 1000 millimeter sticks. Let us suppose, instead, that we have a length meter, as opposed to a 1-meter length, and that it can be applied to objects to yield a reading of length. Applied to a given object on a series of occasions, it gives a series of different readings. If we suppose that a given object *has* a given length (thus treating "length" as in some degree a theoretical function), then the data consisting of length readings, combined with the minimization and distribution principles, can yield a quantitative statistical theory of error.

Regard the readings of length of a given object as having the form $t_v+e_1, t_v+e_2, \ldots t_v+e_n$, where t_v is the true (unknown) length of the object. These data are in the Ur-metacorpus **MK***. Taking statistical inference for granted, we may infer with a probability high enough to warrant inclusion in the evidential metacorpus **MK** that these numbers are *approximately* normally distributed with mean t_v^* and variance s. That is, we have adequate reason to accept that in the total population of such measurements, however large it may be, readings will have a discrete multinomial distribution (since the number of readings is finite) that is *close* to the normal distribution of mean t_v^* and variance s.

It is at this point that the minimization principle would enter if we were deriving our knowledge of error *ab initio*. Of course, knowing what we know, we would not take a small sample to affect our background knowledge concerning the accuracy of a certain method of measurement; what we are doing here is exhibiting the fact that we *could* use our observations themselves to construct a theory of errors of measurement, and that we could do so without circularity.

In the case of a number of measurements of a single object, with no background knowledge concerning errors of measurement, what minimizes the e_i—the errors of observation—is the identification of t_v^*, the mean of the inferred distribution, with t_v the true length. We assume that this distribution of errors of measurement applies to all the measurements with which we are concerned—that there is no statistical basis for distinguishing between different classes of measurements. Thus, given a single reading on a new object, we may infer with evidential certainty that its true length lies within k standard deviations of the reading, where k is chosen to correspond to the level of evidential certainty characteristic of **K**.

Thus armed, we may return to the law of additivity. Writing L for the length-of function and CJ for the collinear juxtaposition operator, and ignoring the distinction between magnitudes and real numbers, the law may be stated:

$$L(CJ(x,y)) = L(x) + L(y).$$

To test it, we may select a pair of objects x and y, measure each, and measure their juxtaposition. Suppose that our three readings are $R(x)$, $R(y)$, and $R(CJ(x,y))$. Obviously, we will not have $R(x) + R(y) = R(CJ(x,y))$, due to the presence of error. But the law of additivity isn't supposed to apply to *readings*, anyway; it's supposed to apply to *true* lengths. So let us choose k so that we have the conjunction "$L(x) = R(x) \pm ks \& L(y) = R(y) \pm ks \& L(CJ(x,y)) = R(CJ(x,y)) \pm ks$" in our evidential corpus. (For example, it is ordinarily "incredible" that a measurement should be in error by more than three standard deviations.)

Shades of Popper! This conjunction cannot verify, and cannot even be regarded as an instantiation of, our law, but it *can* refute it. It could be the case that there is no triple of lengths, $L(x)$, $L(y)$, and $L(CJ(x,y))$ that satisfies this conjunction of assertions. That we didn't observe such a case, though we might have, might be taken as supporting the law. That attitude is not Popperian, of course. What we should do is put the law to a *severe* test.

So let us consider a sequence of these tests. If the sequence is long enough, we are almost certain to encounter a test that refutes the law of additivity. This may be seen as follows. Suppose that the law is true and that our error theory is correct. We have selected k so that (say, 999 times out of 1000) the conjunction representing our test is true. But the probability is also very high that in a long run of trials the conjunction will turn out to be false at least once. (The probability is $.9999 = 1 - .999^{10,000}$ that in 10,000 tests, at least 1 will turn out wrong!)

What we do have is a high statistical probability in our practical corpus that the length reading of a juxtaposition will be close to the sum of the length of the components. This is handy, but we would like more.

Consider, then, adding to the conventions of our language; specifically, let us add the constraint on the juxtaposition operator and the length function that the length of a juxtaposition is the sum of the lengths of its components. Take this new constraint to be an a priori truth. How do our bodies of knowledge change?

Our previous derivation of the statistical error distribution for length readings no longer applies. We cannot simply identify the mean of a set of readings with the true length of an object, since it will not in general be the case that the mean of the readings of $CJ(x,y)$ will be the sum of the means of the readings of x and y. But subject to the constraint that $L(CJ(x,y)) = L(x) + L(y)$, we may still apply the minimum error principle (and now we also need the distribution principle) to obtain in **MK*** statistical knowledge of the distribution of errors of length readings. This distribution will be characterized by a larger variance than before. If we adopt the language in which the additivity of length is an a priori principle, we must accept that length readings are subject to larger errors than if we don't adopt that principle. This represents loss of predictive information; for example, we will have $L(x) = 3.000 \pm .012$ in our corpus of practical certainties rather than $L(x) = 3.000 \pm .010$.

But we also gain predictive information because we can now use that principle, even in the evidential corpus, to derive new statements.

Suppose that we can accept in the evidential corpus the statement "$z = CJ(x,y)$," as we may on either alternative. In the metacorpus **MK$_1$*** of the first alternative we have (say) $O(X, "z = CJ(x,y)", t)$, $O(X, "R(x) = 3.000", t)$, and $O(X, "R(y) = 5.000", t)$, where "$R(x) = 3.000$" means that the result of measuring x yielded the observation report 3.000. Through the known error distribution of individual measurements we may get "$L(x) = 3.000 \pm .010$" and "$L(x) = 5.000 \pm .010$" in the evidential corpus **K$_1$**. But through the accumulation of length observations in **MK$_1$*** we will also have a statement in **MK$_1$** reflecting the correlation between measurements of the lengths of juxtapositions and the sum of the measurements of the lengths of their components. So we will know *something* in the evidential corpus about the length of the juxtaposition of x and y, say, "$L(z) = 8.000 \pm .100$."

On the second alternative, though we may only have "$L(x) = 3.000 \pm .012$" and "$L(y) = 5.000 \pm .012$" in the evidential corpus, we will *also* have (in virtue of the independence of the measurements of x and y) $L(x) + L(y) = 8.000 \pm (2)^{1/2}.012$," and therefore, in view of the additivity of length, "$L(z) = 8.000 \pm (2)^{1/2}.012$." We lose a little in our knowledge of the lengths of x and y but gain a lot in our knowledge of the length of z. (Note that we also have "$L(x) + L(y) = 8.000 \pm (2)^{1/2}.012$" in $\mathbf{L}_1$, but lacking the additivity law, this does not give us the length of z.)

Of course, this benign result need not ensue. It would not happen if we were measuring intelligence and considering the additive law that the intelligence of a committee was the sum of the intelligences of its members. In that case, incorporating the additive constraint would require that we attribute so much error to our observations that we would have little content left in our evidential corpus, a result we would richly deserve.

How can we measure the gain in content provided by the law of additivity? As before, add a finite number of observation reports to $\mathbf{MK}_1^*$ and $\mathbf{MI}_2^*$. Both $\mathbf{K}_1$ and $\mathbf{K}_2$ will contain length statements concerning the terms that appear in these reports, asserting that the length of x lies in the interval I. The content of such an assertion is greater if the interval is shorter. The language with the additivity law is epistemically preferable to the language without the additivity law, just in case the sum of the lengths of the new length intervals mentioned in the corpus of practical certainties corresponding to the language with additivity is less than the sum of the new length intervals mentioned in the practical corpus of the other language.

But we get more. For example, hypothetical observations of the lengths of x, y, and z may yield in the evidential corpus the sentence that z is not the collinear juxtaposition of x and y. (If the length of x is between 2 and 3, and that of y is between 4 and 5, and that of z is between 9 and 10, all with *joint* evidential certainty, it is evidentially certain that z is not the juxtaposition of x and y.)

We must accept the uncertainty of the observation sentences "$CJ(x,y)$", "$Rigidbody(x)$", and so on. But clearly, under the right circumstances — given an appropriate body of *prima facie* observations in $\mathbf{MK}^*$ — the constraints represented by the additivity law will not seriously undermine either our ability to tell rigid bodies and collinear juxtapositions when we see them or our ability to make reliable measurements.

In short, although the question of how to judge the predictive observational content of the corpus of practical certainties is not at all trivial, it is the sort of question that seems to represent a relatively reasonable technical problem.

"Empirical" Laws

It certainly isn't very shocking, whatever one's persuasion about the analysis of concepts may be, to be told that the law of additivity of lengths can be construed as functioning as an a priori constraint on our scientific language. Geometric laws fall into the same category: the area of a rectangle is the product of its height and width. Is this not just a geometric truth? Well, yes, in Euclidean geometry, but it is not without empirical import, since we can measure length and area independently. (It is only by picking units of length and area that are appropriately related that we get the law in the simple form just stated.) The same things are true of the relation of velocity, distance, and time, and of various other physical quantities we think of as being introduced "by definition": most of these quantities admit of direct measurement (though perhaps only in a rather crude way), and so in most cases there is some empirical import to the equation corresponding to the official "definition," just as there is some empirical import to the law of additivity of lengths.

We could regard such laws as these as "empirical," but in view of the fact that there are people who would regard them as analytic as well as a priori, they are perhaps not ideal laws with which to illustrate the grounds of choosing between conventions. Let us consider instead the law of linear thermal expansion. Let us leave to one side the fact that it isn't quite true (there are second- and third-order terms in engineering handbooks) that it depends on constant pressure, and that the measurement of temperature often employs this very law.

The law, applied to a particular substance, has this form:

$$L(x,t_1) - L(x,t_2) = k(t_1 - t_2)L(x,t_1)$$

The change of length of x is proportional to its original length at t_1 and the temperature change it has undergone, where the constant of proportionality is characteristic of the substance involved.

What is the raw material for getting this law? A set of temperature measurements and a set of length measurements. But these measurements are subject to error. If we took the error distributions to be normal, the error would be unbounded, and no set of measurements would be *inconsistent* with the law. Even a realistic set of distributions of measurement error would not allow us to "refute" the law by observation. What we need to do is to regard the observations as recorded in the metacorpus **MK***: "$O(X,"length_1(x)",t) \& O(X,"temp_1(x)",t)$", and so on. From the known error distributions for the measurement of length and temperature, we may now include in the evidential corpus such statements as "$tt_1(x) \in temp_1(x) \pm d$," $tl_1(x) \in length_1(x) \pm f$," and so on, where "*tt*" and "*tl*" stand for true temperature and true length. It takes two sets of such statements to compute k—that is, to find an interval $[k_1, k_2]$ in which, *assuming the law to be true*, it is almost certain that k is in.

We require not merely that each of the four measurement statements involved be evidentially certain, but that their conjunction be evidentially certain. There is no problem in principle here, nor is there any problem about deriving an interval in which we can be evidentially certain that k lies. But, of course, these measurements don't *test* the law.

To test the law, we need more measurements. So let us take a third pair of measurements. It is quite possible, just as in the case of our test of the law of the additivity of length, that the values we get will be incompatible with our presumptive linear law. But just as, in that case, there were two ways of looking at this potential incompatibility — either as suggesting the rejection of the additive law or as throwing new light on the distribution of errors of measurement of length — so in this case there are several ways of looking at the possible incompatibility.

We may regard it as (a) casting doubt on the law of linear thermal expansion, or (b) as providing new evidence concerning the errors of measurement of temperature, or (c) as providing new evidence concerning the errors of measurement of length, or (d) as providing new evidence concerning the value of the (hypothetical) constant k.

Given that all of our measurements are subject to error, there is certainly no categorical refutation of this law. Can we refute it "within the limits of experimental error"? No, because however wide we set the "limits of experimental error," there is a finite (and calculable) chance that the law will be rejected even if it is true. This way of talking suggests that we *can* test the law in the sense of the modern theory of testing statistical hypotheses: we can choose a discrepancy such that, if the law is true, the chance that we will falsely reject it on that basis is small enough to satisfy us.

But that won't quite do, either, since it supposes that the error functions for the measurement of temperature and length are just *given*. They aren't. In the first place, they are inferred from bodies of data, and it is not at all clear that the data provided by our experiments to test the law of thermal expansion should not be included in the data from which the error functions are inferred. In the second place, we have no universal error functions for the measurement of length and temperature, but error functions that differ according to circumstances. We could, that is, infer that temperature and length are particularly difficult to measure reliably under the circumstances that prevail when we are trying to test the expansion law.

Suppose, now, that we regard the law of thermal expansion as a possible linguistic convention: L_2 contains it, L_1 does not. **MK$_1$*** and **MK$_2$*** contain our length readings and temperature readings. Both incorporate whatever a priori linguistic constraints we have imposed on temperatures and lengths. **MK$_2$*** does but **MK$_1$*** does not include the constraint corresponding to the law of thermal expansion. In either case, the constraints give us the error distributions (by means of the minimization and distribution principles) we must attribute to our sample of readings and to various identifiable subsamples of them. These in turn give rise, via statistical inference, to general statistical representations of the distribution of errors in the various populations. These error theories are accepted in **MK$_1$** and **MK$_2$** — the metacorpora of evidential certainty.

Suppose that there is no noticeable difference in the error distributions for the measurement of temperature and length in the two cases. Then, since L_2 gives a predictive mechanism that L^1 lacks, L^2 is preferable. Add observations about the length of a body at one temperature to the metacorpus, and an observation about a new temperature, and then a statement concerning the length of the body at the second temperature will be part of the evidential corpus in L_2. In L_1 we may take advantage of a *correlation* between measurements of temperature and length, but this will not give us such precise knowledge. The situation is the same as that in the case of the additive law for lengths.

Suppose that there is a significant difference in the error distributions obtained under L_1 and L_2. We may distinguish two cases. In the first case, we do not have enough data to distinguish the errors of measurement made in the temperature/length context from the errors of measurement made in other contexts. But since under L_2 the data must be bent to satisfy an additional constraint, the dispersion of the error distributions for both length and temperature will have to be larger than under L_1. Then the question is whether the improvement in knowledge provided by prediction as opposed to correlation is overbalanced by the general increased dispersion or not.

In the second case, we have the standard classical error functions for both measurement and length in both L_1 and L_2 so far as any *other* context but that of thermal expansion is concerned. L_1 has the same functions for thermal expansion it always did. But L_2, which embodies our alleged law, requires special error functions with wide (and perhaps unsymmetric) dispersions for this special context of thermal expansion. It seems clear that in this latter case there will be a loss of predictive content.

Consider something approaching reality: that at a certain degree of crudeness, the linear law is acceptable and useful. We refine our procedures for measuring temperature and length, and we find that the *linear* law can't earn its keep, but that a law with a second-order term both gives us prediction *and* is compatible with our accepted statistics of errors of measurement. We move to the second-order law. Given yet further refinements, we may be driven to a third-order law. (This represents the state of the art as represented in Perry's *Chemical Engineers Handbook*, 1946 edition.)

The upshot is that: suppose that we speak a certain language. Suppose that we consider adding a certain generalization to the language as one of its "rules." We consider the relative advantages of the original and the supplemented language. There are two cases: in the first, we have no statistical basis for distinguishing the errors of measurement of the quantities involved in the generalization depending on whether the measurements are being used for prediction in accord with the generalization or for some other purpose. In that case, we lose some accuracy (thereby decreasing the informational content of our predictions) as the price of the predictions to which the law entitles us. In the second case, we can distinguish (in the language with the generalization) between the statistics of error for observations involving the new generalization and the statistics of error for other observations. If the statistics of error for observations involving the new generalization are particularly bad (as we would ordinarily say: when experiments do not support the generalization), we see that the generalization entails a loss of predictive content in our language in the corpus of practical certainties.

The General Case

Let us apply this approach generally. Construe both the laws and theories of science as *constraints* on our scientific language. Consider the addition and deletion of laws or theories to our scientific corpus and the replacement of one theory by another.

In the case of addition, we suppose that L_2, our new language, includes a statement (or a group of statements) expressing additional constraints on the language L_1: it may also involve new vocabulary. There is a certain class of statements in L_1, observations statements, that are such that X (a person, society, scientific group) may judge directly that a statement of this class holds. Such judgments need not be incorrigible, but they must be such that in many cases they are so little prone to error that they may be accepted in the evidential corpus K of X on the basis of X's observational judgment. The observations are reported in MK_1^*, and the error distributions, derived from the a priori constraints of L_1 and the observations reported in MK_1^* appear in MK_1. (To do this, we employ the minimization and distribution principles.)

The practical corpus K_1' contains all those statements whose probability relative to the evidential corpus K_1 is high enough. It includes statements belonging to the class of observation statements (at least those belonging to K_1 in virtue of observation, but presumably others as well).

Suppose that MK_2^* records exactly the same observational judgments as MK_1^*. Then since L_2 imposes more constraints on the language — fewer sets of sentences are consistent — the sample frequency of error we need to attribute to the observation sentences recorded in MK_2^* must be at least as great as or greater than the sample frequency of error we need to attribute to the observation sentences recorded in MK_1^*. Statistical inference then yields error frequencies for the various kinds of observation statements that are at least as great as those reflected in MK_1. We therefore get less material in K_2 on the basis of observation than we get in K_1, or at any rate, no more. But the new constraints allow us to expand "analytically" what is in K_2 on the basis of observation, and this, in turn, can lead to an expansion of the content of the practical corpus K_2'.

Since what we want to measure is the predictive observational content of K_2', we should hypothetically augment each of MK_1^* and MK_1^* by N statements $O(X, S_i, t)$, where the S_i are observation statements appropriately distributed among the possible *kinds* of observation statements. The appropriate distribution mirrors the historical relative frequencies of these various kinds of statements. If for large enough N, K_2' contains more content than K_1', then the addition to the constraints of the language is warranted. Content is measured (crudely) in terms of the number of basic statements ("Aa," "$\sim Rabc$," etc.) and for quantitative statements in terms of the narrowness of the intervals in which the quantities are asserted to lie.

This, I suggest, accounts for additions to the a priori constraints embodied in our scientific language. How about deletions? The same analysis suffices. If, by the standard just outlined, L_1 is preferable to L_2, we should adopt L_1, lacking the proposed additional constraint of L_2. But this leaves something to say, since we can *delete* a law or theory *only after* we have already *accepted* it. So suppose that with one set of observation reports as a base in MK_2^* and MK_1^*, L_2 is preferable to L_1. Is it plausible to suppose that an increase in the number of observational reports serving as a basis in MK_1^* and MK_2^* for our error statistics can lead to a reversal in that preference? If so, how?

Recall that the error distributions in MK_1 and MK_2 are determined, via statistical inference, from the numbers of various sorts of observation statements in MK_1^* and MK_2^* that must be rejected to achieve consistency with the axioms of L_1 and L_2, respectively. The most obvious monkey wrench that can be thrown into our preference for L_2 over L_1 would be apparently direct counterexamples to the law or theory. It would not take many of them to establish that there was a *class* of observations in which the relative frequency of error had to be taken to be excessively large. But whether this leads to the preferability of L_1 over L_2 depends on other factors as well. If the power of L_2 to generate predictive observational statements in K_2' is very great, we may be better off, even in the face of apparent counterexamples (the orbit of Mercury) to stick to L_2, and to hope that some further modification of our language, or some new boundary condition, will allow us once again to treat observations of that special class as accurate. It is a question of a trade-off, and the terms of the trade-off lie, I claim, in the contents of the corpus of practical certainties.

In general, of course, we do not observe even *apparent* direct counterexamples to laws and theories. What happens is that through some chain of reasoning, uncertain in every link, we derive a class of predictions that don't come off. As Duhem pointed out in detail, it takes a whole theory of physics to test a special and limited law of physics.

Again, however, we can be led to attribute unacceptable error to a particular class of observations, and this can lead to an impoverishment of our corpus of practical certainties, and therefore to the preferability of L_1 over L_2. It need not, of course. It may be, as in Duhem's examples,[3] that there are a number of ways of saving the accuracy of our measurements. Our standard for preference, in this case, even gives us some guidance as to which way is preferable: that deletion that maximizes the hypothetical predictive observational content of our corpus of practical certainties is preferable.

The replacement of one theory by another could, of course, be construed as a deletion followed by an addition, as Levi[4] construes it. (This is also the direction in which the logic of theory replacement has developed.[5]) But it seems more realistic (and historical) to consider the replacement as a contest between two languages, L_1 and L_2, where L_1 embodies one constraint and L_2 another. By the criterion advertised above, we can compare these two languages directly in terms of the hypothetical predictive content of the corresponding practical corpora. In fact, one can go further: it may well be the case that both L_1 and L_2 are preferable to the language L_0 that embodies neither of the competing constraints. Thus the move from L_1 to L_0 would be unwarranted. Levi struggles with the fact that this move would result in a "loss of information." Without judging the success of his struggle, I note that in the present framework it is unnecessary.

There is another form of scientific change that is so far largely unremarked, except by those whose emphasis has not been on the

---

[3]Pierre Duhem, *The Aim and Structure of Physical Theory*, Princeton University Press, Princeton, N.J., 1954.

[4]Isaac Levi, *The Enterprise of Knowledge*, MIT Press, Cambridge, Mass., 1980.

[5]For example, see Carlos E. Alchourron and David Makinson, "On the Logic of Theory Change," *Theoria* **48**, 1982, 14–37. The most formal and complete exposition is Carlos Alchourron, Peter Gardenfors, and David Makinson, "On the Logic of Theory Change: Partial Meet Contraction and Revision Functions," *Journal of Symbolic Logic* **50**, 1985, 510–530. All this work accepts Levi's proposal that theory replacement involves contraction followed by expansion.

rationality of such changes.[6] This is the change in the observation vocabulary itself. By this I mean nothing so radical as the claim that Ptolemy "observed" the sun's rising while Copernicus "observed" the rotation of the earth. I have in mind Galileo and his successors learning to observe heavenly bodies through a telescope, and Leowenhoek and his successors learning to observe cells through a microscope. Note that this is still a significant part of medical education: learning to see what you are "supposed" to see through a microscope.

The change in observation vocabulary entails a difference between the metacorpora MK_1^* and MK_2^* containing observation reports. A difference between these metacorpora would also be generated by a change in X's observational vocabulary even without the change being due to scientific novelty. For example, being a Polynesian, X might move to Alaska and learn to distinguish the 13 varieties of snow distinguished by Eskimos, meanwhile losing his ability (after many years) to distinguish the 17 varieties of waves he used to be able to distinguish.

In general, though, we obtain new abilities to make observational judgments without losing old ones. Whether it is worth the effort or not depends on the resulting increase in the predictive observational content of our practical corpus. There is little doubt that learning to identify cells under a microscope is epistemically worthwhile. We may compare this with learning to identify the lumps on a head or the lines in a palm. These, too, require practice and training by experts. And it is surely possible that people with the requisite practice and training will make, almost always, the same observation reports, while those without the training will simply "not see what is there." So far, the phrenologist is right: the amateur phrenologist, like the amateur histologist, is not in a position to challenge the professional's observation.

The question is, what can you do with it when you get it? There may indeed (I suppose) be phrenological theories, which, if adopted as constraints on our language, would lead to predictive consequences. But now you see the rub: these theories provide a connection between phrenological observations and more ordinary observations (of aggressiveness, of success in business or love, etc.), but these very connec-

[6]Examples: Norwood Russell Hanson, *Perception and Discovery*, Freeman Cooper, San Francisco, 1969; Thomas S. Kuhn, *The Structure of Scientific Revolutions*, University of Chicago Press, Chicago, 1962; Paul Feyerabend, "Problems of Empiricism," in R. G. Colodny (ed.), *Beyond the Edge of Certainty*, University of Pittsburgh Press, Pittsburgh, 1965, pp. 145–260.

tions may undermine the dependability, the reliability, of both our phrenological observations and our ordinary observations. If the error rates get too high, we will have succeeded only in impoverishing our practical corpus.

Note, however, that it takes a significant number of observation reports of a given kind in **MK*** in order to derive a useful statistical statement of error frequency in **MK**$_1$. Thus it is not unreasonable that it should take some time for new methods of observation to become established. Galileo could see the moons of Jupiter when someone else could only see spots of light, some being (in fact) moons of Jupiter, some being defects of the optical apparatus, and some being perceptual anomalies.

The other side of the coin is that it also takes some time for new methods of observation to become *dis*established. It is not immediately and a priori obvious that spots under the microscope are therapeutically important and that lumps on the head are not. The moral is that one should keep an open mind, but that one shouldn't keep it *wide* open for an *indefinite* length of time.

What, then, of the semantic content of our theories? It is true that there *are* bacteria and that they are causally tied to human infirmities. It is false that the lumps on a head fall into a certain structured set of classes and that these classes are causally tied to human character traits. Or so we firmly and rationally believe. It has been suggested (e.g., by van Fraassen[7]) that what we *should* believe is that the world behaves *as if* our favorite scientific theories are true. That is, that these theories serve as useful *instruments* for getting around in the world and manipulating it, but cannot claim ontological or metaphysical truth. This view is known as "instrumentalism." The alternative, that science gives us the truth about the world, is known as "realism."

This contrast between realism and instrumentalism is not easy to formulate in the framework we are considering.[8] It would be quite unnatural nowadays to deny the reality of what we can see with our own eyes under a microscope: that is observable, and the contrast between realism and instrumentalism concerns the theoretical. But in

[7]See Bas van Fraassen, *The Scientific Image*, Oxford University Press, 1980, Oxford.

[8]It is not easy to formulate in any framework. It can be stated as a metaphysical distinction or as a linguistic distinction. It can be regarded as a question of what the truth status of theories is or as a question of what entities there are in the world. And it can be thought of as a question of whether the principle of bivalence (every statement is true or false) is a matter of logic or acceptable at all. And there are other variations.

the days of Leowenhoek it was not so clear that what one "saw" through the tube were *objects*. Practice with other optical apparatus could make it easier to see these things as objects—continuity plays a large role in these matters. It is not clear that concern with the contrast between realism and instrumentalism is as enlightening as the focus on errors of observation and on the predictive observational content of the practical corpus.

You will observe that there is no question about the truth of the theory of scientific inference and change that we have examined. Since we've argued that all theories are true a priori (being linguistic conventions), the same is true of *this* theory: it couldn't be *false*, since it is true a priori. Of course, you can argue that some other conventions, or fewer conventions, would be preferable, and it may be that you can do so successfully by exhibiting the resulting increased contents of our joint corpus of practical certainties. But wouldn't that show that the present theory was right after all?

9

Relativity and Revolution

Observability

In the last chapter, we considered in very general terms what can happen when you cut the ties between observability and certainty. More needs to be said, however, before we can apply the framework we developed there to the kinds of real theories we are interested in: relativity, quantum mechanics, and other such highbrow creations. In particular, we need to look more closely at the nature of those statements in the corpus of practical certainties that constitute the predictive observational content of that corpus that we get from the sorts of theories we really do accept, and we need to consider the choice among distinct theories, if such there be, having the same predictive observational content.

A sentence in the corpus of practical certainties of level p' is part of its predictive observational content only if it is the *sort* of sentence that we can include in our evidential corpus on the basis of observational judgment, but that does not occur there on that basis. That is, we want to focus on statements in $\mathbf{K}'$ that have *not* gotten there on the basis of observation but that might have. Clearly, any statement rendered probable enough by observation to occur in $\mathbf{K}$ will also have been rendered probable enough by observation to occur in $\mathbf{K}'$, so these statements also are not part of the predictive observational content.

The corpus of practical certainties contains, by inheritance, statements corresponding to mathematical and set-theoretical truths, implications of the axioms of our theories and the like, none of which are to be counted as part of the predictive observational content of the corpus. We wish to count as part of the predictive observational content only basic sentences — sentences consisting of a predicate, followed by an appropriate number of terms, or the negation of such a statement, that is of the sort that can come to be accepted on the basis of observational judgment. Note that the cash value of measurement statements—the length of A lies between 10 and 12 feet—is embodied in other statements that are generated by them, and in particular in the set of " . . . is longer than A" and "A is longer than . . . " statements.

Such a sentence need not *be* confirmed by observation, of course, and in fact, we may know perfectly well that it will not be. We may never be in a position, or put ourselves in a position, to test it by observation. We may be practically certain that there is a paper in the drawer, and never feel moved to open it; practically certain that the sugar in the bowl is soluble, and never be moved to dissolve it; practically certain that the ladder is between 10 and 12 feet long, and never be in a position to measure it.

Consider a sentence "$Ra_1a_2 \ldots a_n$" belonging to the predictive observational content of **K'**. For it to be *observational*, it must be the case that if an observation report containing that sentence were to appear in **MK***, then the corresponding observation sentence would justifiably appear in **K**. Observational judgments of this form must almost never need to be rejected—"almost never" being cashed out in terms of what we take to be the level of evidential certainty. For it to be *predictive*, it must be the case that it does *not* already appear among the observation reports of **MK***.

It is natural enough to think in terms of the reliability with which the predicates of a language can be applied, but we must also think of the terms $a_1, \ldots, a_n$ to which they are applied. What might "a_1" be? "The next A to be observed by me" is a plausible candidate, in which case it is as dependable as A-observations in general. Suppose that "a_1" is more complicated—for example, "the seventeenth item that is A and B but not C, to pass between D and E after 2:43:54 A.M., 12/15/88." Could one become evidentially certain that some object satisfied this description? There seems to be no reason why not. One might have to take rather extraordinary measures, and even be a bit lucky. One might need a troop of Boy Scouts, as dedicated to truthfulness as to cleanliness and godliness combined, acute of eye, and judiciously located; one might need careful coordination with the National Bureau of Standards to keep track of the time. But, given p, the level of evidential certainty, it should, even in the case of complicated definite descriptions (but not necessarily "tricky" ones), to establish with error frequency less than $1-p$ that a certain object answers to the description "a_1."

So the descriptions may be complicated, particularly if, as surely must, we take science to be a collective and social enterprise. Note that while we may or may not have proper names in our language, we will be able to accept, on the basis of observation judgment, statements involving definite descriptions. Since these statements can be unpacked into quantifications, to say this is to say that we may accept certain quantified judgments on the basis of observation: the x such that A is P.

Unpacked, we have: there is an x such that A, and for every y, if y is A, then $y=x$. This seems like a lot to swallow: can we really get a universal generalization from observation? But this is not so unreasonable; as Russell pointed out long ago, we can surely confirm the universal generalization "there are no lions in the garden" by observation.

The sentences in the predictive observational content of our corpus of practical certainties that interest us the most are those that pertain to measurement. From the (perhaps hypothetical) results of measurements $m_1, m_2, \ldots, m_j$, with the help of theory T, we infer that a certain quantity lies between q_1 and q_2 with practical certainty. From this, in turn, we infer that the object in question (to which the quantity applies) bears the relation R to such and such other objects, where R is the relation underlying the quantitative measure. But there are various pairs q_1, q_2 such that we can, with our given evidential corpus, be practically certain that the quantity in question lies between q_1 and q_2. We must, in measuring the predictive observational content of $\mathbf{K}'$, focus on a single pair.

The principle by which we do so has already been employed in talking of the acceptance of statistical hypotheses. This same problem arises in classical confidence interval analysis and is resolved by focusing, where possible, on the *shortest* confidence intervals. Note that this is what we are already doing in binomial inference, as previously described, and that "shortest" intervals are "most informative" intervals; the principle could be motivated by considerations of information maximization.

The same principle is used when we use our knowledge of an error distribution to pass from an observed value of a measurement to an interval of acceptance for the corresponding quantity. Although the error may fall in the interval $[-.010, +10.00]$ with the same probability as it may fall in the interval $[-1, +1]$, it is the latter rather than the former that we use (under ordinary circumstances) to determine an acceptable range for the quantity being measured.

Let us look more closely at a quantitative prediction. Suppose that it followed from (perhaps hypothetical) measurements together with a quantitative theory that the length of x lies between b Q-units and c Q-units. Now consider the quantity $L(x)$ in the expression $bQ < L(x) < cQ$. How do we know that it satisfies those inequalities? It is as if we had measured it. And indeed, we have. For it follows from no theory or piece of science by itself that x should have a length in that range. It might follow from a direct measurement of x; but it might also follow from other measurements together with some theoretical relationships—that is, from an *indirect* measurement of x. The numbers b and c, then, are chosen precisely to yield a shortest interval for $L(x)$ such that the inequality is probable enough to be accepted as a practical (or evidential) certainty.

It is from this shortest inequality that we derive the predictive observational inequalities relating x to other objects. Thus, if we have $b'Q < L(x') < c'Q$ and $c < b'$, then we should have the observational sentence "x' is longer than x." In particular, if x' is that segment of a meter stick (or whatever) for which $L(x') = b$ Q-units, and x'' is the segment such that $L(x'') = c$ Q-units, we should have "longer than (x, x')" and "longer than (x'', x)," which is equivalent to the result of a direct measurement observation: I report the result of measuring x as $L(x) = d$ Q-units; from which could follow with practical certainty: "longer than (x, x')" and "longer than (x'', x)."

Now let us consider the other question: what if we have distinct theories that lead, given our observations, to a corpus of practical certainties having the same predictive observational content? Or is this possible at all?

Let us begin by considering cases of identity between two scientific theories embodied in different languages. Clearly, if the only difference between the two languages is that one uses "F" where the other uses "G," and vice versa, we have no significant difference. This is a matter of what Grünbaum has called "trivial semantic conventionalism."[1]

Almost as trivial is the case of a change of unit. It seems intuitively that it should make no difference in our theory if we measure the

---

[1] Adolf Grunbaum, *Geometry and Chronometry in Philosophical Perspective*, University of Minnesota Press, Minneapolis, 1968, p. 5.

distance in inches or meters. If determinate values of quantities are construed as magnitudes, as we construed them earlier, then indeed, this is merely an instance of trivial semantic conventionalism: 39.7 k inches *is* exactly the same (abstract) object as k meters. If quantities are construed as numbers, it isn't quite so clear: is the *property* of having a value of the length-in-inches function equal to 39.7 k the same *property* as that of having a value of the length-in-meters function equal to k? If it isn't, it should be.[2]

But there are distinct theories that have the same predictive observational consequences. Although it would be painful to exhibit instances, we could; it is easier to establish the existence of such theories directly. Suppose that we are given the sentences in the evidential corpus (consequences, we assume, of observation reports in the Urmetacorpus). Relative to this set of sentences, the set of sentences comprising the predictive observational content of the practical corpus can be (we have been supposing all along) characterized recursively. If that is the case, we can accept recursive constraints (on a language that has only the observational vocabulary) that are exactly such that the given observations lead to the set of predictive observational certainties in the corpus of practical certainties.

But no more in this case than in the case of the original version of Craig's lemma[3] do we want to say that the more primitive language is preferable to, or even on a par with, the more natural theoretical language. Why not?

One natural answer is that it is a matter of computational complexity; but this answer (even if true) leaves two things open: Why is it that we should measure computational complexity in one way rather than another? And why should we take computational simplicity as a guide to truth and a guide to what there is? What is the connection between truth and computational simplicity? I am not sure what the answers to these questions are. But I do suggest that there is a different question

---

[2]In applications of the so-called pi theorem of dimensional analysis, it may seem that we get substantive empirical results from the consideration that our units shouldn't matter. But this is an illusion. For a clear discussion of this matter, see R. L. Causey, "Derived Measurement, Dimensions, and Dimensional Analysis," *Philosophy of Science* **36**, 1969, 252–270.

[3]William Craig, "Replacement of Auxiliary Expressions," *Philosophical Review* **65**, 1956, 38–55. See also C. G. Hempel, "The Theoretician's Dilemma," in H. Feigl, M. Scriven, and G. Maxwell (eds.), *Minnesota Studies in the Philosophy of Science*, Vol. 2, University of Minnesota Press, Minneapolis, 1958, pp. 37–98.

that might be asked, namely, why should we accept computational simplicity as a guide to credibility or credence-worthiness?

The answer to that question I offer with some trepidation. I'm not even sure how persuasive it is. But if our main goal is to make predictions, to manipulate our environment to satisfy our own desires, then surely our *goals* are best attained by whatever procedure attains them *and* is computationally most economical. Since our goals are equally attained by two theories that have the same predictive observational consequences (given our observational input), the computationally most economical one should be preferred. Does that mean that the entities, and so on, of the preferred theory really exist? Does it mean that we should believe that they exist? What more should we want than this last? What more justification for believing the pronouncements of contemporary science — if they *are* instrumentally justified — do we need?

Note that the arguments (or considerations) offered here are not arguments for instrumentalism. I am not suggesting that fancy theories are only justified as instrumental devices for making predictions. Rather, I am suggesting that fancy theories can be justified by means of their instrumental uses. Why not, instead of believing that the world is Φ, believe that the world is as if Φ? Because this involves one extra computational step: from the data to the imaginary world to the prediction.

To put flesh on the bones of this suggested argument would require considerable attention to detail. But it seems to me that a reasonable case can be made for it.

Relativity

We have seen that universal generalizations involving observational predicates can be regarded as embodying a priori features of the language we use, and therefore, in a sense, should be construed as "conventional." But we have also seen that we can have epistemic grounds, depending on the course of our experience of the world, for choosing between a language that incorporates a particular generalization and one that does not. Similarly, it has been claimed that the additive law for lengths, which many consider "analytic" of the very concept of length, is indeed an a priori feature of our language, but one that is justified by our experience. It might have been the case that to adopt the additivity of length in our language would have required the attribution of so much error to our measurements of length that we would have suffered a net loss to our corpus of practical certainties. Similar things can be said about the addition of other elementary quantitative laws to our corpus of knowledge—the law of linear thermal expansion, Hooke's law, and so on.

There is an air of artificiality about all these examples. It is due to the fact that to illustrate how these conventions function in our rational corpora, and the way in which we can have epistemic justification for adopting them, we tend to speak of very sparse and simplified bodies of knowledge. In this chapter, we shall try to show that the approach we are exploring need not get totally bogged down in complex detail, even if we concern ourselves with relatively far-reaching and pervasive theories.

As Quine is fond of pointing out, our scientific knowledge is all of a piece. To use the metaphor of Quine and Ullian,[4] it is a web of interlocking and connected strands, supporting each other, and attached to reality only at the edges by means of observation sentences. By introducing observational error explicitly, I have (somewhat) cut the web loose at its edges. But I have also allowed observation to penetrate the interior of the web: any term we can learn to apply directly with high reliability can be regarded as an "observation" term. Histologists, in view of their experience, can observe mitochondria; the laymen cannot.

It could be argued that histologists are "inferring" the presence of mitochondria from the colored patterns on the slide and from a flock of facts about optics and microscopy, but this is of a piece with the claim (to which even Quine sometimes appears prone) that we infer the presence of cats and dogs from the irritations of our retinas. There is no law against using the word "infer" in this way, but I am giving it a more natural and commonsense sense. I take *argument* as fundamental; observation starts where argument ends. It is then that the question of the reliability of a particular sort of alleged observation can be raised.

This seems straightforward enough; in fact, it seems Pickwickian to say that in these simple situations of accepting generalizations, we are "changing our whole language." What needs examination is the case where the new theory, for example the theory of special relativity,

---

[4]W. V. O. Quine and J. S. Ullian, *The Web of Belief*, Random House, New York, 1970.

changes the meanings of the terms (as embodied in the constraints they are assumed to satisfy) that are pervasive in the language (e.g., the function "mass of"). Not only does the new theory affect the interpretation of "mass of" as applied to planets and photons, but in virtue of the connection between mass and weight, it affects the local butcher and Mrs. Jones' motivation for dieting. Some people—Feyerabend, for example—go so far as to say that Newtonian theory and special relativity can't even contradict each other because their terms (like "mass of"), though spelled the same way, have different meanings.

On the face of it, this is an odd thing to say. For, speaking loosely, Newtonian theory says that the black spot representing star X will appear on the photographic negative at place p, and relativity says that it will appear at place p', where p and p' are just different places. Surely the two theories in question have "spot" and "place" as common parts of their vocabularies.

But this may be too fast. Newtonian theory does not predict anything about spots on negatives directly. It is only relative to some assumptions, some boundary conditions, that such predictions can be made, and even then, what emerges may be construed as a matter, not of certainty, but of probability. So it might be thought that to compare Newtonian theory with special relativity, we should look at relatively "theoretical" assertions, and such assertions would (presumably) concern distances, masses, velocities, energies, and the like. At the same time, however, from a down-to-earth, practical point of view, the difference lies in what the two theories can do for us in the realm of practical certainties and decisions.

Both of these aspects of the comparison of the two theories can be captured in our framework. Let us look at the case in somewhat more detail. The claim that special relativity and Newtonian theory attribute different meanings to the function "mass of" is true, on the present view, at least to the extent that "mass of" is subject to different constraints in the languages corresponding to the two theories. But, having located our knowledge of the world in the corpus of practical certainties, let us see how far the changes extend.

First, in both the corpus of practical certainties and the evidential corpus, there are a vast number of nomic generalizations and statistical generalizations that do not involve mass (or any other suspect term) at all. If we decide that it is to be a feature of our language that all crows are black—that is, if the generalization earns its keep, as we spelled that out earlier—that is unaffected by the controversy at the foundations of physics. When we compute the death rate from cancer on the basis of epidemiological data, and when we use that rate to compute the probability that a policy holder in an insurance company will die from cancer, we do not need to change a thing when we change the constraints that "mass of" satisfies.[5] Well-made dice yield 6s a sixth of the time today, just as they did in 1800.

Second, there are the butchers and bakers and engineers. Weighing objects, and the distribution of errors in such weighings, will be essentially unchanged in view of two facts: One, such weighings are done at relatively low velocities, and thus "theoretically" there is relatively little difference to be expected in the error functions. Two, the source of our knowledge about errors of measurement is generally directly empirical. For example, we might weigh the same object a number of times and use those weighings as (part of) the basis for a statistical hypothesis about the distribution of error in weighings by that apparatus. Or we might calibrate our scale against a standard.

Here is where the idea that our corpus of practical certainties contains approximate as well as uncertain (though practically certain) statements is important. If our practical corpus contained such statements as "The mass of this ball bearing is π grams," then of course, in virtue of the fact that it is moving with some velocity relative to the earth, we would have to attribute to it a new mass. But what appears in our corpus is only the claim that it weighs between g and g' grams, and although there may exist a number of significant figures at which this is true for Newtonian mass and false for Einsteinian mass, that number is so large that the difference does not concern us.

A strong case can be made that the numbers g and g', representing the limits within which it is rational for us to believe that the mass of the ball bearing lies, literally do not change. The data on the basis of which we derive our error distribution are recorded to only a finite number of significant figures. We can argue that the error distribution is discrete, with no more cells than correspond to this number of significant figures. We do not have arbitrary precision in the error distribution, and therefore cannot have arbitrary precision in the end points of the intervals in which quantities are reasonably believed to lie.

[5]Oops! Cancer is sometimes tied to radiation, which is tied to the degree to which we are willing to act on the basis of relativistic theories, since it is they that underlie the convertibility of mass and energy. No special relativity, no nuclear power plants, less cancer (maybe).

So, ordinary scientific statements that do not involve mass don't change. Most ordinary engineering statements concerning mass don't change. Ordinary statements—such as the butcher's—about weight don't change. But there are some statements regarding mass that do change: statements involving masses that may be moving at high velocities, statements involving the relation of mass and energy, statements involving quite ordinary-sized masses of heavy elements (such as plutonium) in small spatial volumes.

There are a few statements in the corpus of practical certainties that are strikingly different. Included in the predictive observational content of the relativistic corpus of practical certainties may be such statements as "If button A is pressed, then city B will burn up."

But this sort of change in the predictive observational content of the corpus of practical certainties is well down the road. What changes initially? Very little that we can put to the test. Traditionally, we have the bending of light rays near the sun and the account of the precession of the perihelion of Mercury. There are, of course, all sorts of predictive statements we could get into our hypothetical rational corpus—that the mass of Old Bess, our Percheron, who is hypothetically traveling with a velocity close to that of the speed of light (though in fact she only moves at a slow walk), is much greater than her normal mass. Even statements about the conversion of mass into energy are among the new statements. But all of these statements are matched one for one by statements of the Newtonian theory: Bess's mass at high velocities is the same as her mass in her stall, there is no conversion of mass into energy, and so on. It requires high scientific genius to find hypothetical changes in the evidential corpus that we can make *actual* and that yield significant changes in the corpus of practical certainties.

Testing Relativity

Let us focus, for the moment, on the novel consequences of the theory of special relativity that could, with the technology available at the time it was put forth, be put to the test. The precession of the perihelion of Mercury could not be accounted for by Newtonian theory; it was an anomaly. With regard to all other celestial measurements, a certain standard theory of error was reasonably believed to be appropriate. Measurements pertaining to the orbit of Mercury did not fit.

Given the Newtonian language, these measurements were not random members of the general set of celestial measurements with respect to embodying certain amounts of error, but rather were a special case. We had to regard the measurements as being subject to peculiar and unusual (unexplained) errors, or to regard the deviation of the orbit of Mercury as a result of some "unknown" cause.

Einstein's proposal brought the data concerning the orbit of mercury into conformity with other celestial data only in the sense that the same error distribution applied to observations of Mercury that applied to other planets without the introduction of ad hoc hypotheses. But Einstein's proposal carried with it implications that are at least as unnatural and strange as many of the hypotheses that had already been proposed (e.g., the existence of the dark planet Vulcan) to account for the peculiarity of Mercury's behavior. Note that since Mercury's orbit was already known, there was nothing *new* for Einstein's theory to add to our practical corpus in this case. We already knew where Mercury would be next June, because we preferred to think that there was something odd about the parameters of Mercury, as opposed to our measurements of other celestial objects. What we didn't understand was why it should be *there* rather than where Newtonian theory said it should be.

The case of the bending of light rays near the sun is quite different. This is often alleged to be a classical case of a crucial experiment: Newtonian theory predicted one thing, Einstein's another, and all we had to do, according to the simple story, is to see which is correct and which is refuted. Matters are not so simple. First, consider the case in which we had no alternative theory; the results of the Eddington[6] expedition would clearly not have been regarded as refuting Newtonian theory. Suppose that we had Einstein, but no Newton. The results of the Eddington expedition would have been quite unnecessary; the behavior of the tides, the solar system, and so on would have been quite sufficient to have shown the usefulness of the Einsteinian framework. And, of course, in neither case would we have been motivated to perform such a difficult and expensive experiment. The experiment is *interesting* only when we have both alternatives before us.

---

[6]The Eddington expedition traveled to Africa (1904) to take advantage of a total eclipse of the sun to measure the distortion of light as it passed close to the sun's surface. Incidentally, Newtonian theory does not predict no distortion — just a smaller amount than does Einstein's.

We have been thinking of the two competing theories as conventions. This is easy to do, since they are relatively abstract and removed from experience. The connection between the world of experience and the application of either theory is in any case, only probablistic. Not only does this render either theory immune to strict refutation, but it is the theory combined with observations that gives us the distribution of errors of measurement that allow us to connect the theory with experience even probabilistically.

But the whole theory is not needed for this. In fact, with regard to the general error distributions relevant to measuring the apparent angles of celestial bodies with the help of telescopes, Newton and Einstein give us exactly (well within any conceivable margin of error) the same results.

Now let us return to Africa. The results of our measurements are a set of observation reports in our metacorpus. These are combined and processed through our statistical theory of error to yield intervals in our evidential corpus. We can be quite sure that the angle α that interests us lies between d_1 and d_2 degrees.

Now consider what I am calling the two languages. In the Newtonian corpus, prior to the observations, we have the prediction (in our corpus of practical certainties) that α lies between n_1 and n_2. In the Einsteinian corpus, correspondingly, we have the prediction that α lies between e_1 and e_2. Note that the prediction is approximate and occurs only in the corpus of practical certainties, since the (quantitative) boundary conditions are fixed only approximately and probabilistically in the evidential corpus. Furthermore, we may suppose (at least in this case) that they are derived from the same observation reports.

Let us represent the quantitative conditions on which the α prediction is based (in either language) as

$$V \in R$$

Roughly speaking, a certain vector of quantities lies in a certain region. The size of this region depends, of course, on the theory (as well as on what we are taking to be the level of evidential certainty).

Let us see what happens in the two cases we have been considering when we make the crucial observations. Suppose that A is the body of data on which the error distribution for measurements of angles (under the conditions at issue) is based. In the *absence* of the constraints imposed by the two theories (leading to the predictions for α), we would expect to be able to accept with evidential certainty that α lies in the interval $[d_1,d_2]$. But we do have these constraints. Let B be the background data (presumed to be relatively independent of both Einsteinian and Newtonian general theories) relative to which the margins of error in the two predictions ($\alpha \in [n_1,n_2]$ and $\alpha \in [e_1,e_2]$) are established.

Once we have made our observations, the source of the error distribution by means of which we pass from the observation reports concerning α to the observation statements concerning α must be taken to be the combination of A and B. The result of this is not independent of which language we adopt. Suppose that $[e_1,e_2]$ and $[d_1,d_2]$ overlap a lot and that $[n_1,n_2]$ and $[d_1,d_2]$ do not. Then the combination of the error data A and the error data B, in the relativistic framework, will yield a resultant interval $[ed_1,ed_2]$ that is neither much broader than nor very different from the "predicted" interval $[e_1,e_2]$. The combination of A and B in the Newtonian framework, on the other hand, will yield a resultant interval $[nd_1,nd_2]$ that is considerably wider than the original interval $[n_1,n_2]$.

In other words, if, after the experiment, we ask how much knowledge we have of α (the quantity measured), we can answer: much more in the relativistic system than in the Newtonian system. Furthermore, if we suppose—we will consider grounds for such suppositions in a later chapter—that similar experiments would yield similar results, then hypothetical Newtonian and relativistic corpora would be characterized by the same asymmetry of information.

The conventional description of this situation is that the observations are consistent with the relativistic theory and not with Newtonian theory. This option is not open to us, since, in taking error seriously, we have made the observations consistent with *both* theories. It is true that on each theory we can calculate the probability, in advance of our observations of α, that α will be found in the interval $[d_1,d_2]$. But of course, we know that that is the interval we should consider only *after* we have performed our experiment. What we learn then is that finding an α observation in $[d_1,d_2]$ is much more improbable in the Newtonian framework than in the relativistic framework.

This in itself does not cast doubt on Newton's convention—the improbable happens. Every bridge hand is extremely unlikely. Under Einstein's conventions, the observation—that is, the data interval—is not so improbable. But so what? We could invent a theory (after the event) according to which the probability is $[1,1]$ that α will be reported in the interval $[d_1,d_2]$. What we should be concerned with, after the experiment, is what the true value of α is. Both theories give an answer to the effect that the true value of α lies in a certain interval; but the answer from the combined data from relativity theory is more precise, more informative, than the answer from Newtonian theory.

Revolution and Convention

The best-known view concerning the replacement of one theory by another is Thomas Kuhn's view[7] that the replacement is revolutionary. Along with this we often find the suggestion that the replacement is arbitrary, a sociological event rather than a rational decision. Kuhn's view is that science changes (one hesitates to say "progresses") by saltations; a branch of science chugs along under one paradigm, producing useful but rather pedestrian "results," until some genius, impelled by an increasing weight of difficulties (anomalies) in the tradition, restructures the whole language of that branch of science and thereby generates a new paradigm.

This surely doesn't seem right. We like to think that the changes constitute progress — an augmentation of our knowledge about the world. Although Kuhn himself continues to insist that the historical replacement of one theory by another does constitute progress, and that there *may* be rational grounds for such a replacement, these claims seem more like pious testimony than reasoned conclusions.

There is something reminiscent of Kuhn's view in the view we are considering, according to which a change of theory should be regarded as a change of language. It is clear that it makes no sense to say that the evidence supports a certain language to a certain degree when the evidence must be written in that very language. And the discontinuity remarked by Kuhn is reflected in the fact that major changes of scientific language are relatively rare. In general, science does chug along under one paradigm (one language) until some genius restructures the whole language of that branch of science and thereby creates a new way of talking and writing and experimenting. Novel generalizations are rare.

It seems less plausible and natural to suppose that *every* time a universal generalization is added to our store of knowledge, that represents a change of our language of science. It can be claimed, however,

---

[7]Thomas Kuhn, *The Structure of Scientific Revolutions*, University of Chicago Press, Chicago, 1962. A discussion of this point of view, and a defense by Kuhn, may be found in I. Lakatos and A. Musgrave (eds.), *Criticism and the Growth of Knowledge*, Cambridge University Press, Cambridge, 1970.

in partial response to this observation, that universal generalizations are not often added to our store of knowledge.

As Popper[8] has never tired of pointing out, the wonderful thing about universal generalizations is that they can be refuted; but one cannot "refute" a generalization that is a built-in part of the language.[9] That particular objection was dealt with earlier, in Chapter 4 and, interestingly, applies as well to Kuhn's view as to mine. In part, my response to the objection that we do not think of our scientific language as changing all the time (though, of course, it changes when new vocabulary is added—not a rare event) is that despite the tendency of philosophers to focus on universal generalizations in science, these universal generalizations are rather rare birds.

Even in ordinary language, we use universal generalizations to express what are more properly approximate statistical generalizations with parameters close to 0 or 1. We say that all crows are black or that all swans are white, but we don't really believe it. And evidence that we don't really believe it is provided by our reluctance to convert our generalizations (to argue from " all crows are black" and "X is not black" to "X is not a crow"), our propensity to cite only the evidence provided by crows in supporting the alleged "universal generalization," and so on.

Another prevalent view is one that could be associated with Quine and Duhem. According to this view, our body of scientific knowledge is tied to reality only around the edges; the center of the theory, the part removed from observation, is highly underdetermined. There are many ways our body of knowledge could be that are all equally compatible not only with the experience we have had so far, but with the totality of an infinite body of experience.

The popular thesis of underdetermination is supported by the web metaphor. If the web is tied to reality only at its edges, then any old interior with the same edges is as good as any other, barring nonepistemic criteria such as simplicity and elegance. We can change the knots and strands at a certain place in the interior of the web in whatever way we please, provided that we make appropriate adjustments elsewhere.

The suggestion that the generalizations, laws, and theories of science should be regarded as a priori features of the scientific language

---

[8]K. R. Popper, *The Logic of Scientific Discovery*, Hutchinson, London, 1959. (First German edition, 1934.)

[9]See Popper's contribution and Kuhn's reply in Lakatos and Musgrave, op. cit.

we speak seems to support a similar holism. What is up for grabs when we consider a new theory is the whole language of science. This seems extreme, particularly as applied to simple generalizations. But as we have already noted, universal generalizations are not as common in our bodies of knowledge as philosophers seem to think. Furthermore, if one of these common-variety universal generalizations is challenged ("Are you *sure* that all crows are black?"), the response is often to return with a blatantly linguistic version of the generalization: "Well, all *normal* crows!"

From the present perspective, we can accommodate both this intuition that science progresses by saltations, and that large advances occur relatively rarely, and a kind of holism of the sort supported by Duhem or Quine. When we are epistemically warranted in changing our language by the incorporation of a universal generalization as an a priori feature, it is because we thereby increase our knowledge about the world. More precisely, we have increased the (predictive) content of our corpus of practical certainties. (To those philosophers who say that this corpus cannot be construed as knowledge because not all of the statements in it are — or perhaps can be — true, we may reply, "So much the worse for your construal of 'knowledge'.")

Note that in the case of a simple universal generalization, the adjustment to our language is minimal and localized. In our example of A's and B's, only the terms A and B and their relation were affected. We changed the reliability with which we could apply the terms A, B, $\sim A$, and $\sim B$, but not by much. We added the constraint that all A's are B's. This might have been as far as the changes extended. They might also have extended somewhat further, for example, if our language already contained the statement that all B's are C's. The adoption of "all A's are B's" would then have had an effect on the reliability we attribute to C-judgments.

Things are somewhat more complex in the case of more general and far-reaching theories, but not much more. As is clear from our discussion of Newtonian and relativistic mechanics, there are vast regions — even regions using the terms "mass," "velocity," and "energy" — of our corpus of practical certainties that are quite unaffected by our change of language. The change of constraints governing the term "mass" has no bearing on the question of whether I should diet or not. The revolutions, even in the most extreme cases, are less pervasive than we have been led to believe.

10

Idealization

Idealization and Approximation

Consider the test of some quantitative law—say, the law of linear thermal expansion. Of course, as we have already remarked and as everyone knows, the test results do not exactly fit the law. But we put the discrepancy down to errors of measurement, and in fact, we may often use those discrepancies to provide evidence for the distribution of errors in that context.

Sometimes, however, we can go beyond that. Suppose that the test concerns the construction of a thermometer. What would we standardize the thermometer against? We would have to have a procedure for constructing standard temperature reservoirs. We might, for example, take one part water at the temperature of melting ice and four parts at the temperature of vaporizing steam to obtain a standard temperature that we might call "80." Of course, the objective validity of this procedure depends on the fact that the specific heat of water does not vary significantly between 0° and 100°C. The epistemic validity of the procedure depends only on the fact that we do not observe (judge) that the specific heat changes.

It is possible that the errors are not distributed in the same way at different points on the scale: at 20 the mean error may be positive; at 40 it may be negative. We might say, "*Ideally*, the error distribution would be the same for any point on the scale." There are two things we might mean by this. One is merely that it would be "nice" if the error distribution were uniform. It is in this sense that the ill-informed legislature of a midwestern state once passed a law to the effect that π should have the integral value 3; ideally, the ratio of the circumference to the diameter of a circle would be 3.0. But this is not a sense that will concern us.

A more serious thing we might say is that if the thermometer were perfectly made (perfectly uniform bore, infinitely more mercury in the bulb than in the tube, etc.) and perfectly used (infinitely carefully?), then there would be no errors. This is more serious, because it reflects actual knowledge of the measuring process: the relevance of the uniformity of the bore, the ratio of the amount of mercury in the bulb to the amount of mercury in the tube, and so on. And perhaps we can always be a little more careful about how we use the thermometer (use magnifying devices, avoid parallax, etc.). But there are two difficulties of opposite sorts with this sense of "ideally."

First, of course, is that we cannot make a bore perfectly uniform; we cannot have an infinite ratio between the amount of mercury in the bulb and in the tube; we cannot be infinitely perceptive. So it is not clear that this conception of ideality is any more helpful than the previous conception.

Second is the fact that if we *were* able to accomplish these impossibilities, we would have to regard the original claim as false, for it would *not* be the case that the errors would be uniformly distributed: the expansion of mercury (or anything else) is not really linear. There are, as we know, higher-order terms.

We need a conception of ideality that allows us to be more realistic. Something like this might do: *if* the thermometer had been made more carefully, *then* the distribution of error at various points on the scale would be *more* uniform.[1] This requires that we have some idea of an explanation for the discrepancy and some idea of how to reduce it. It is an essentially *comparative* conception.

There are several features of this comparative notion of ideality that we should note here for future reference. One is that ideality can be approached, but not necessarily arbitrarily closely. In our thermometer example, as we improve the physical thermometer, we ap-

---

[1] This blatantly counterfactual conditional is not as hard to unpack in our extensional framework as one might think. Assume that the error at any point t on our temperature scale is (roughly) normally distributed, with a mean error of $e(t)$. The error itself will have some distribution as a function of t. The claim, then, is that if we replace the knowledge we have about the uniformity of the thermometer by (hypothetical) knowledge according to which it is more carefully made, then we will have reason to suppose that we will know that the distribution of $e(t)$ will be more like a uniform distribution.

proach the ideal of linear thermal expansion only for a while. Eventually our thermometer becomes so ideal that it reflects, in biased distributions of error at different temperatures, the fact that the thermal expansion of mercury is not really linear.

In order for an ideal to make sense, it is not necessary, obviously, that we be able to reach it. It is sufficient that we be able to approach it, and even then, only to a certain degree. We need not even be able to approach the ideal arbitrarily closely, even (or especially!) "theoretically."

In order to make sense of the "improvement" we get in approaching an ideal, we must have some measure of how close to the ideal we are. In the case of the thermometer, the measure was given by the bias in the distribution of error at different points on the scale. We had an essentially internal measure, a measure related directly to the relevant kind of measurement. In other cases, we might have a more external measure — for example, in using light interference patterns to determine the extent of our approach to an ideally flat surface.

We should also note that there may be more than one respect in which we can approach the ideal, and that it need not be possible to approach the ideal in all of these respects at once. Thus an ideal balance is arbitrarily rigid and arbitrarily light; but at some point, long before we get to the ideal, any increase in rigidity requires an increase in mass, and any decrease in mass entails a decrease in rigidity.

The ideality that we have been talking about concerns the ideality of the physical thermometer. We have not considered the ideality of the thermometric substance. And yet that fits into the same pattern. We cannot make a given substance more ideal for the measurement of temperature, but we can *observe* differences in the degree to which different substances are suited for the measurement of temperature. "Suitability" must be measured in a number of different dimensions — something else that should be noted about ideality in general. For example, the ideal thermometric substance is fluid over a wide range of temperatures, does not undergo significant changes of specific heat over a wide range of temperatures and so on.

But so far as our prior concerns go, what we can compare are the degrees to which the expansion of various liquids at the temperatures with which we are concerned are truly linear. None of them is really linear, but we can *imagine* an ideal thermometric substance whose expansion is really linear.

Ideal Laws

As a more interesting example of idealization, consider the ideal gas law, $PV=nRT$, where P is pressure, V is volume, and T is temperature. The constant n characterizes the quantity of gas. As in the case of the additivity law for length, if we adopt a Popperian stance, we can be very sure that even if the law is true, we will be able to falsify it. But worse, it is the *ideal* gas law, and so we know that it doesn't quite hold anyway. As we can discover from a statistical analysis of the actual measurements made in connection with the law, there are *systematic* deviations.

Suppose that our measurements of pressure and volume are not to be changed and that n (the number of moles of the gas) is taken as given. (We leave aside the fact that one way of getting at the molecular weight of a gaseous compound is by way of the ideal gas law.) R is the gas constant, the same for all gases. We can evaluate it (within limits, with evidential certainty) given the distribution of errors in our measurements of temperature, pressure, volume, and quantity. The dispersion of measurements of R depends on the particular gas involved, but the lower the pressure, the higher the temperature, and the lower the molecular weight of the gas, the better behaved these measurements are.

Should we accept the ideal gas law as a conventional a priori feature of our language? Since there are no ideal gases, we needn't worry about its being falsified; we already know that it is false of any real gas. But what *good* does it do us if there is nothing to which it applies?

First, it applies *approximately* to a wide range of gases under a wide range of circumstances. Thus it can be used in the same way that the law of linear thermometric expansion discussed in the previous section can be used.

Second, we can distinguish the circumstances under which it applies with greater or lesser degrees of approximation, and we can specify the direction in which the behavior of a given gas will deviate from that of the ideal gas. This is reminiscent of Pierre Duhem's emphasis on increasingly close approximations as guiding the direction of scientific research.[2]

---

[2] Pierre Duhem, *The Aim and Structure of Physical Theory*, Princeton University Press, Princeton, N.J., 1954.

Third, we know what to do in many circumstances to bring the behavior of actual gases into closer conformity with the hypothetical behavior of an ideal gas.

So far, we could construe the ideal gas law merely as a handy rule of thumb. It could be taken as a guide to the distributions of measurements of gases under various circumstances.

But the name suggests more. The law may also be construed as an a priori constraint both on our use of the phrase "ideal gas" and on our measurements of pressure, volume, and temperature. Although the constraint tells us that there are no ideal gases, it allows us to distinguish on independent grounds instances that are closer to or farther from ideality. I have assumed that we know all about measuring pressure and volume, and the distribution of errors of such measurements. To assume the same for temperature would be less plausible. In fact, at high temperatures, it is the ideal gas law itself that helps us determine the distribution of errors of measurement of temperature. This fits in with the fact that for actual gases, we take the approximation provided by the ideal gas law to be better at high temperatures and low densities. We have good theoretical grounds for this now, though before the victory of the atomic conception of matter, the story was different. The ideal gas law thus functions not only as an approximative ideal but as a normative ideal: *accurate* measurements of temperature, under good conditions, are those that are close to being in conformity with it.

What is the epistemic advantage conferred on us by the use of a language containing an ideal quantitative law that applies to nothing? One might answer: simply the epistemic advantage conferred on us by the procedures of science. In effect, that is the answer offered by Nancy Cartwright in her aptly entitled book, *How the Laws of Physics Lie*.[3] But we should look at the reasons a bit more closely.

For one thing, even if the ideal gas law were quite literally and ideally true, it would add nothing to the contents of our corpus of practical certainties all by itself. It requires evidential input to provide any output. So suppose that the time is now and that sample is our sample of gas. If we measure the number of moles of this gas in the sample (by weighing it and knowing the atomic weight of the gas), and measure its temperature, and measure its volume, we can predict what its pressure must be. Of course, all of these measurements admit of

---

[3] Nancy Cartwright, *How the Laws of Physics Lie*, Clarendon Press, Oxford, 1983.

error; so the prediction that can be part of the content of our corpus of practical certainties is not that the pressure is (exactly) so and so, but that it falls in a certain interval.

In treating the ideal law as an approximation to reality, the same kind of thing occurs. Rather than concluding that the pressure is within the interval, we conclude that it is approximately within that interval—that is, that it *is* within a wider interval. Of course, in special cases, we can do better: knowing something about the gas in the sample, we can know whether the ideal gas law will yield a prediction erring above or erring below the true value of the pressure; knowing exactly what gas is involved can lead us to an even more precise prediction of the pressure—though never more precise than the original ideal interval, since that imprecision is derived from measurement error alone.

The prediction of a pressure is conveniently simple. More complex predictions should be taken account of, too. Knowing how many moles of gas comprise the sample, and knowing its temperature, we can predict the product of the pressure and the volumes; knowing pressure and volume and temperature, we can predict n, the number of moles. And, of course, even if we know nothing about the sample except that it is a gas, we can predict that it will approximately satisfy the ideal gas law.

Note, however, that in order to obtain a categorical prediction from the ideal gas law, we must accept something as evidentially certain: we do not regard the conditional, "if this sample is a sample of gas, then it more or less satisfies the ideal gas law," as part of the predictive observational content of our practical corpus.

The ideal gas law contributes to more than just the predictive observational content of our practical corpus, however. That contribution would warrant its incorporation into our language at some point in the development of our thermodynamic knowledge, but it would become superseded by substance-specific knowledge of the *PVT* relations: when you have steam tables, you don't use the ideal gas law to compute the properties of steam.

What is this "more"? There is the way in which the ideal gas law interacts with other laws to yield predictive content. This is a matter we will explore more thoroughly in the following chapter. There is also the fact that to regard it as a law is to regard it in some sense as "true in the limit." There are several glosses that may put on the phrase "true in the limit."

1. We may be able to specify ideal circumstances under which the law would hold exactly. Thus (to anticipate) we can give the exact Newtonian solution to the two-body problem. This is the problem of how two isolated bodies, with given masses and initial positions and velocities, would continue to move. In the corpus of knowledge that incorporates Newtonian mechanics, this solution holds *exactly* in a universe consisting of just the two bodies. The ideal gas law does not admit of a limiting case of this sort.
2. We may be able to specify a way of approaching, arbitrarily closely, a case in which the law applies exactly. Thus the behavior of the two-body system approximates more and more closely to the Newtonian standard, the farther removed from all other bodies is the system in question. For a real gas, the lower the pressure, the larger the volume, and the higher the temperature, the more nearly does the gas obey the law in question. This does not embody quite the same sense of approximation involved in the first mechanical case. There was no constraint there on the masses of the bodies or their distances, but here we are imposing constraints on the values of the variables of the law itself.
3. We may not be able to approach ideality arbitrarily closely (even in principle, even for special values of the variables), but we may be able to *account for* the failures of ideality. This is the case for the real applications of the ideal gas law. So long as the interactions of the molecules of gas are small compared to the kinetic energy of those molecules, and we have a way of estimating the effect of these interactions, the gas will act relatively ideally. When large pressures or low temperatures force the particles of the gas into circumstances where the intermolecular forces and molecular volumes play a significant role, then there will be significant departures from ideality. But we may still know how to take account of these departures.

There is a further dimension along which the approach to ideality can take place; that is, gases can be ordered with respect to how well they conform to the ideal gas law. Gases of low molecular weight do better. But this does not provide us with a pattern of approximation that we can extend—at least, not easily. It does give us further (ultimately predictive) information about the nature of gases and compounds.

Ideal Boundary Conditions

In applying Newtonian theory to the solar system, it is customary to make some simplifying assumptions. The celestial bodies are treated as point masses (even those that are *internally* complex, such as the subsystem of Jupiter and its moons). In addition, we (initially) take each planet (treated as a point mass) together with the sun (treated as a point mass) as a two-body system (i.e., we assume that the influence of other celestial bodies is negligible). Then we do the same for all other pairs of bodies, and then we combine the results of these idealized computations (treating the initial pairs as point masses), and so on. What is amazing is that this procedure often works (to a reasonable approximation) and, furthermore, that the approximation can be improved on by iterations of this same procedure.

There is another way in which this sort of idealization plays a role in the development of physical theory. The general equations of motion, for example, may involve an unspecified force function. In any particular application, we may have to provide the particular form of the force function (and then experiment to evaluate constants, and then measure boundary conditions, and then infer outcomes in order to obtain a prediction). We often arrive at the appropriate form of the force function by considering an *idealized* version of the problem under consideration.

To use an elementary example, one can infer Snell's law (the ratio between the sine of the angle of incidence and the sine of the angle of refraction is constant for a given pair of media) from quite abstract conservation principles, subject to the assumption that the two media are isotropic. Since media in the real world tend to be anisotropic, this is a mere idealization. But it is a useful one, since many media come *close* to being isotropic, and the law is good for making reasonably accurate predictions concerning such media. Note that what is idealized here is not the fundamental law—the conservation principle—but the *boundary condition* by means of which this law is applied to a particular class of phenomena.

This procedure is so pervasive in engineering practice that it almost escapes notice by its familiarity. A man carries a pole on his shoulder with a weight at each end. We consider the man a point, the pole a rigid body of zero cross section, and the weights as point masses. By so doing, we get a solution to an idealized problem that has no counterpart in the real world. But the idealization is such that the predictions made in accordance with it appropriately appear as part of the predictive observational content of our corpus of practical certainties.

More interesting examples occur in more sophisticated contexts. As we noted earlier, there are often several respects in which a certain situation may be idealized, and it may not be possible to approach them all simultaneously. A more realistic model of the man with a pole on his shoulder might treat the pole as rigid, but not as weightless, or it might treat the pole as weightless but attempt to take account of its elasticity. In the case of quantum mechanics, we learn from Nancy Cartwright,[4] it is sometimes the case that it is only the result of trying out different idealizations that leads us to a decision concerning which is appropriate in a given context.

"Shall we treat a CW gas laser below threshold as a 'narrowband black body source' rather than the 'quieted stabilized oscillator' that models it above threshold? Quantum theory does not answer. But once we have decided to describe it as a narrowband black body source, the principles of the theory tell what equations will govern it. At the first stage [of theory entry] there are not theoretical principles at all—only rules of thumb and the prospect of a good prediction."[5]

The sorts of things, Cartwright points out, that one studies in studying quantum mechanics could hardly be more idealized: the free particle in one dimension, the particle in a box, piecewise constant potentials and so on. This is a matter, not of the practical subengineering computation of the stress on a man's shoulder, but of fundamental theory. Before one even measures, one commits oneself to idealization and therefore to approximation.

A particularly interesting example, from our point of view, is provided by statistical mechanics, though the conclusions I draw from the successes of statistical mechanics differ from those drawn by Cartwright, who writes, "I do not think these distributions are real. . . . In the vast majority of these [situations] it is incredible to think that there is a true probability distribution for that situation."[6]

From the point of view adopted here, these distributions are certainly "real": there is, for a given sample of gas at a given time, an exact proportion of the molecules that have a velocity less than or equal to v. But the distribution we pick to serve as a basis for charac-

---

[4]Ibid., p. 46.
[5]Ibid., p. 134.
[6]Ibid., p. 136.

terizing the situation abstractly, or for making predictions, is almost certainly *not* the distribution that really obtains. We pick the distribution we do in part because it is one that we can deal with mathematically — it is "tractable" — and in part because it corresponds to an ideal state of affairs. How can we get away with such a thing? Because the ideal here, as elsewhere, is *close* to the actual.

The distribution we use is very close to the actual distribution in the precise sense that the first few moments of that distribution, on which our predictions depend, are very close to the first few moments of the actual distribution. In the aspects of the distribution that matter, the ideal distribution is a perfectly adequate surrogate for the actual distribution.

Fundamental Idealization

It has long been regarded as puzzling, at best, and a refutation of the classical treatment of scientific theory, at worst, that the Newtonian laws contradict themselves. Well, almost contradict themselves. One law states that any body not acted upon by external forces will continue in its state of rest or rectilinear motion; and the accompanying gravitational theory says (in effect) that all bodies are acted upon by forces. In first-order logic, then, any conditional whose antecedent is "x is a body that is not acted upon by any force" will be true simply by virtue of the falsity of the antecedent. "All bodies not acted upon by external force will do figure eights" would be just as true as the first law.

One way to deal with this situation is to emphasize the counterfactual character of the first law; it may be read as "If a body *were* not acted upon . . . " and that statement, in turn, may be interpreted nontruth-functionally — that is, so that the falsity of the antecedent does not entail the truth of the whole conditional.

Another approach is to construe it as an *ideal* law along the lines of the ideal gas law. It is true, it could be admitted, that no bodies are free of external forces, and we all know that; but there are bodies that are, in various respects, *relatively* free of external forces. And in this form, the law not only says something but provides us with predictive content of the most common and everyday sort. For example, when a trailer breaks loose from a car going down a highway, the trailer will continue in rectilinear motion even if the car and highway turn. But, of course, the trailer is acted upon by external forces: for example, the force of gravity holding it to the surface of the earth (and attracting it to the sun); and its motion is not really rectilinear: it curves with the surface of the earth, corkscrews with the earth's spin, and follows the earth's trajectory about the sun. None of this matters: these are not the forces and rectilinearities with which we are concerned. There is an element of idealization in this example with which we are concerned: namely, the retarding force of friction and, ultimately, the retarding forces provided by fences, trees, and other vehicles. But by removing obstacles, greasing the wheels, and making the road smooth, we can approach the ideal more closely.

This may further our understanding of the function of the first law, but it does not help our formal representation of Newton's laws. While we could write (roughly) "For all x, if x is an ideal gas, then $P(x)V(x) = nRT(x)$," we cannot write Newton's laws for an ideal body. But we can write them for an ideal *system* of bodies — that is, one that is isolated (in relevant respects) from the rest of the universe. So written, the laws apply additively: it is their collective import that represents the motions of the bodies in the system. By increasing the isolation of the system (by dealing with the solar system, rather than merely with the Earth and its moon), we can improve the match between the real and the ideal.

What then of the *sentence* that says that "Any body not acted on by external forces . . . "? Therein lies the genius, or part of the genius, of Newton's approach. By itself, that sentence indeed says nothing — or not much — of scientific interest. This is so even if we construe it counterfactually: if the universe *were* to consist of a single particle, that particle would not be acted upon by external forces, and so *would* continue in its state of rest or rectilinear motion forever. Who cares?

It is only in combination with the rest of the system that the first law makes its contribution. We can therefore represent the Newtonian theory in our body of knowledge as a quantified set of open sentences with roughly the sense, "For any F and S, if S is an isolated system of bodies and F is the set of forces acting on and among them, then their spatiotemporal behavior evolves according to such and such formulas," where the curious first law is represented among formulas.

We do not have here merely an approximate law: the law is intended to be exact and to apply exactly to actual bodies. Furthermore, we are not merely dealing with boundary conditions — though, of course, that is going on when we apply the law to actual bodies. It is our inaccuracies and approximations in the application of the law that account for any deviations between the law and the world, according to the Newtonian believer. It is the law itself that is the ideal.

Other Ideals

Idealization, it should be clear, plays an indispensable role in the physical sciences. But the use of idealizations in the physical sciences seems rather straightforward and uncontroversial. We have a good grasp of the conditions under which we can regard a rod as rigid, a system of bodies as isolated, a person as a point mass.

In the social, psychological, and philosophical sciences, idealizations live a much more exciting life. It is not at all clear what the conditions are under which we can regard a person as a "rational agent" or as an "economic man." While in the physical sciences we can find circumstances or create circumstances under which an ideal condition (freedom from external forces, absolute zero temperature, etc.) is very nearly met, it is not at all easy to find uncontroversial instances of this sort of thing in the social sciences.

The responses to this situation in the social sciences have gone in two directions. One way of dealing with the problem is to develop richer theories that are capable of handling the deviations from the ideal circumstances. Actual people deviate from the economic ideal in such and such (ideal!) ways, due to such and such conditions; so we enrich the model in order to take account of these deviations. With bad luck, this might be like seeking a direct analytic solution to the three-body problem; no such solution is known even now. With good luck, it would be more like taking account of the ratio of the length to the diameter in analyzing the mechanical behavior of a column: when that ratio is over about 16, a whole new group of factors increase suddenly in importance.

The other direction is to seek ways of finding or creating unusually uncontaminated circumstances—unusually isolated bodies or rods that are in some important respect highly rigid. This direction is often pursued in psychology, where one can often isolate one's subjects in appropriate ways.

Another related issue concerns normative idealizations. An example is provided by decision theory. The principle of maximizing expected utility can be construed as an ideal descriptive principle that can be used to help predict behavior; or it can be construed as an ideal normative principle, a guide to be followed insofar as one can in one's own decision making.

The expected utility of an act consists of the sum, over all the possible outcomes of that act, of the product of the probability of that outcome, given that act, and the utility of that outcome, given that act. The principle of maximizing expected utility, construed normatively, is that one ought to perform the act that maximizes one's expected utility. Construed descriptively, the principle says that people do, so far as their finitude, time, and computational capacity allow, act so as to maximize their expected utilities.

Even stated thus simply, we are making important idealizations. We are supposing that both utility functions and probability functions are point valued, so that expected utility can be represented by a single number. In the case of probability, we have already seen that this is a nontrivial idealization. Similar remarks could be made regarding utility. And we are assuming that the expected utility of an act is an appropriate representation of its desirability (as opposed, for example, to its most favorable possible outcome or its worst possible outcome).

Of course, as soon as one deals with intervals of probability or intervals of utility, the principle of maximizing expected utility must be reformulated in one way or another. The simplest reformulation is to the effect that a choice that is dominated by another—that is, that has a smaller expectation under *any* permissible probability distribution— should be (is) rejected.

People being finite creatures, there may also be a difference between what they take to be the probability of an outcome conditional on an act and the true probability (relative to what they know) of that outcome conditional on that act. The normative principle, of course, deals with the true probabilities. But the descriptive principle may be taken to concern perceived probabilities. Thus the person who *thinks* that the probabilities are such that act A maximizes her expectation, even if she is wrong about that, is instantiating the empirical principle by choosing act A.

Similarly, there may well be some principles according to which a certain roughly interval-valued utility function is in itself indefensible; yet if an individual chooses in such a way that *if* it were his utility function, *then* he would be maximizing his expected utility by choosing as he does, then he is instantiating the empirical principle. In neither case would we want to bestow the honorific adjective "rational" on the person's choice, even though the person is, in a sense, following the normative ideal principle of maximizing expected utility.

There is another dimension to this maze of normative, descriptive, and ideal considerations that should not go unremarked, since it is fairly pervasive in the social sciences. In order to discuss the principle in question, we have noted, we must have some approximation to a probability function and some approximation to a utility function. Where do these come from? In normative terms, there is no problem for the probability function. It is determined exactly by the content of the evidential corpus of the individual. That is, it is probabilities relative to this corpus that should enter into the person's choice behavior. But in empirical terms, what counts is what the person *thinks* the probabilities are; and how do we, the experimenters, determine this? One way (perhaps, if we count all its variations, the only way) is by looking at the person's choice behavior! The same may be said of utility. But clearly, we cannot simply say that the principle of maximizing expected utility should be accepted as an empirical principle if it is possible to find rough utilities and rough probabilities in such a way that it is roughly instantiated by people's behavior. This can *always* be done.

But now we see the closing of the circle. For this is just the conventionalist's observation concerning the law of the lever, the conservation of mass-energy, dynamics, or whatever: any observations admit of error; so any observations can be construed as consistent with the law in question. On the other hand, in both the physical and the social cases, we can differentiate among laws offered as ideals, as standards, according to their usefulness in generating predictive observational consequences in extensions of the bodies of knowledge we actually have. And it is exactly here — where we have known it was all along — that the big differences between physical and social sciences show up. The best laws in the social sciences give us relatively little, in our present state of knowledge (of boundary conditions, etc.), in the way of predictive observational content.

11
Causality

Cause and Effect

The relation between cause and effect has been regarded as essential to scientific thinking. David Hume devotes many pages of the *Enquiry*[1] to the discussion of causality and appears to take it to be of central importance to our understanding of the world, despite the fact that he can find nothing to the notion, in the final analysis, but constant conjunction. One senses, in Hume's prose, both disappointment and heroic resignation.

But where and when did this notion of causality that Hume sought to understand enter philosophy? I am not sure. Hume's notion of cause *may* be the same as Aristotle's notion of efficient cause, but I am not even sure of that. In the intervening centuries it received a widely differing array of interpretations. What is clear is that with Kant causality becomes both capitalized and elevated to a fundamental category, and that from that time on, it has played a large role in the analysis of scientific inquiry. Mill, for example, regards science as the seeking of causes, and his methods are intended as codifications of the procedures for doing so.[2] Some philosophers regard causality—sometimes even universal causality—as a necessary assumption or basic "presupposition" of science.

It is sometimes claimed that universal causation—or at least its probability[3]—is required for the justification of induction.

---

[1] David Hume, *An Enquiry Concerning Human Understanding*, Open Court, La Salle, Ind., 1949.
[2] John Stuart Mill, *A System of Logic*, Longmans, London, 1949.
[3] This was the view of John Maynard Keynes, expressed in *A Treatise on Probability*, Macmillan, London, 1921.

You will note that our discussion of induction did not so much as use the word "cause". And in fact, since the causal relation is quite clearly an intensional one (one cannot, presumably, replace "A" in "A causes B" by just any old term that is coreferential with A), we have been working all along in a framework that presupposes the unimportance of causality as a relation to be reflected in the object language of science.

But since we have eschewed object language representations of necessity, probability, provability, necessary truth, and the like, without prejudice to the importance of these notions, to bar the causal relation from the object language is not to say that the relation itself is unimportant. Many writers believe that causal explanation is a particularly enlightening form of explanation; clearly, causal relations are important in engineering some events by manipulating others; and according to some writers on decision theory, causal ideas are central to the correct understanding and application of decision theory.

LaPlace, in the preface to his *Theorie Analytique Des Probabilities*,[4] popularizes an image that captures the sense of universal causal determinism: if some intelligence of infinite calculational capacity knew the location and velocity of every particle in the universe and the forces acting on it, at some time, then the whole history of the universe, and the whole future course of the universe, would be fully revealed to him.

In the twentieth century, this view has had some difficult going. Quantum mechanics is irremediably tychistic in its present formulation. Efforts have been made all along, and are still being made, to find a "hidden variable" theory to replace quantum mechanics, so that the tradition of universal causation can still be honored. So far, these efforts have been frustrated.

One response—a response made, for example, by Bertrand Russell in *Human Knowledge*[5]—is to allow that some features of the universe are irreducibly stochastic in character, but to defend a principle of causality nevertheless by allowing that some causal relations are merely statistical or probabilistic. Thus although we cannot predict when a particular atom of radium will emit a particle, we can give an exact, deterministic formula specifying the time rate of decay of radium. It

---

[4]Pierre Simon Marquis de LaPlace, *A Philosophical Essay on Probabilities*, Dover, New York, 1951.

[5]Bertrand Russell, *Human Knowledge*, Simon & Schuster, New York, 1948.

appears that we have merely replaced a law of the form "A always follows B" by a law of the form "A follows B exactly $100p\%$ of the time."

It is not easy to work out the details of such a view in a plausible way. We shall examine some efforts along those lines in the next chapter. For the moment, let it be observed that this is, at any rate, not the view of causation that has been prevalent for many years, and that derives from Kant or earlier writers, and that found its expression in LaPlace's image. According to that notion, there is indeed something very much like a *power* in the world itself or in the events of the world that produces effects from causes. The intuition is that of one billiard ball *smashing* into another and thereby *forcing* it to move in a determinate way.

From a Kantian point of view, we cannot know these powers, but we can only *understand* the world if we see it in such terms. In neither case, though, does the relation of causality obtain between *classes* of events; it is *individual* events that stand (perhaps as individuals of certain types, or under certain descriptions) in the causal relation. And if this is the case, it is hard to see how a statistical law can express a causal relation.

These considerations suggest some distinctions that will be useful in what follows. We should distinguish among determinism, universal causation, and what I shall call, for lack of a better word, uniformity.

Uniformity is the weakest notion; it is reflected in the idea that under the same circumstances the same thing will occur. Perhaps it is intended to be captured by the catch phrase "same cause, same effect." Thus stated, it is, as Russell observed, quite empty of content, since circumstances are never (so far as we know, or so far as we can know) "the same"; at the very least, the date and place have changed. This renders the principle of uniformity true, but in a totally uninteresting way.

It is then suggested that mere difference in time and place should make no difference. But again: there is no mere difference in time (the planets are in different locations; the universe has expanded) or in place (to be in one place rather than another is to bear different spatial relations to every object in the universe). It seems that no ordinary refinement can make something interesting out of uniformity.

Universal causation and determinism are more interesting so far as our interest in science and scientific inference is concerned. But they are not much easier to make precise. To illustrate the distinction between causation and determinism, we can give the following rough characterization: according to determinism, the state of a certain local part of the world is determined by the immediately preceding state of it and its local environment. This might be to say, for example, that there exists a language and a program that will take a description of the immediately preceding state of the local part of the world and its environment as input and give as output a description of the present state of the local part.

This threatens to be as vacuous as the notion of uniformity. Without restrictions on the nature of the program, the richness of the language, and what is to count as "local," the claim is unspecified. But the claim becomes vacuous if we take as properties of the local part of the world its relations to the rest of the universe. Nevertheless, there is presumably some intermediate refinement of language and some specification of locality according to which we could seriously debate whether determinism in this sense is true or not.

Causality is a far more powerful notion. It requires a *connection* of some sort between the earlier and the later states. Not a *logical* connection, as Hume argued so convincingly, but *some* sort of connection in virtue of which an antecedent individual (event) "brings about" a consequent individual. While a deterministic uniformity may be captured by a relation between classes or properties ("crows are black," "combustion is accompanied by heat"), a causal connection requires relations between individuals: *this* fire is the cause of *that* heat; *that* heat is produced by *this* fire. That fire (in general) causes heat (in general) may also be true; it is surely most naturally construed as asserting something about *instances* of fire and of heat.

If this is so, we can see that determinism and (universal) causation are quite independent. We can (conceivably) trace the evolution of a local part of the world from state to state without identifying any causal connections between parts or aspects of one state and parts or aspects of the succeeding state. On the other hand, universal causation may be true, but true in such a way that the evolution of a part of the world is caused in part by relatively remote (nonlocal) circumstances, and in that case we would have to say that determinism is false.

In the sections that follow, we shall consider, first, a notion of causality that seems to be undisputably applicable in the world—namely, the notion of causality that is tied to the notion of agency. We shall then explore the meaning of deterministic uniformity in classical physics and the question of whether (or in what degree) that notion can be (or should be) made universal. Then we will take a closer look at the question of how determinism and causality enter into both explanation and decision.

Agency and Causality

One context in which the notion of causality seems perfectly clear and reasonable and uncontroversial is that in which it serves to point to *responsibility*. "It was the driver of car A that caused the crash" clearly pins responsibility on that person. What is conveyed is that, had he acted in some other way, which he could have, the crash would have been avoided. The same is true, of course, of the driver of the other car: had she acted in some other way — that is, had she left the house 10 minutes earlier, or had she followed another route, the crash would have been avoided. So the claim that had the first driver done something different the crash would have been avoided is not *all* that is conveyed by the assertion that he caused the crash.

What is conveyed in addition is responsibility *as an agent*: blame (or praise) for bringing something about. To build a fire or a bridge, to chop down a tree, or to plant a field of corn is to intervene causally — that is, to manipulate the world. And to know in individual cases that someone has built a fire or a bridge is to understand those cases. I take these to be paradigmatic cases both of the causal relation and of explanation. Causality is obviously tied to engineering and decision making. The presence of fire is explained to me by my knowledge of its cause: it was built by my neighbor. This homely and familiar world, I suggest, is the source of conceptions of both causality and explanation.

The example of the automobile accident has more to offer us. The driver of car A, we say, caused the accident. But the chances are that he caused the accident by driving too fast; it was his excessive speed that was the immediate cause of the accident, and it is he who is responsible for (the causal agent of) his excessive speed. Or perhaps he didn't even intend to drive so fast; it was just a matter of leaving the car under the command of the cruise control when he started down the exit ramp.

What we see from this is that even in the homely example, we can trace a causal sequence or causal chain, only one link of which need be intended by the agent. The driver intends to drive very fast; by so doing, he causes, is responsible for, an accident. The woodsman intends to warm himself; in order to achieve this end, he finds wood, rubs two sticks together, and builds a fire, causing himself to become warm. He is responsible for all the events in the sequence and for the forest fire that results when his camp fire gets out of control. The fundamental locution seems to be: X caused Y by doing Z, or, more generally, X caused $Y_1, Y_2, \ldots, Y_n$, by doing Z, where the Y's are a sequence of things for which X is responsible.

The notion of responsibility, in turn, is easy to generalize. If it isn't your fault, and it isn't his fault, and it certainly isn't my fault, it must nevertheless be somebody's fault. *Someone* must take responsibility. If not us, and not someone in our clan, then someone in another clan; if not any person, then some nonperson — some nonnatural agent. God did it!

Observe that this extension fits in with both the engineering and legal uses of causality. By establishing the cause of the crash, we have fixed responsibility. If no person, then an act of God. We have still fixed responsibility. And just as an ordinary human agent may cause a forest fire by his carelessness, and just as the forest fire is *explained* by the story of the careless camper, so God may cause a forest fire in his anger, and his anger may explain the forest fire. Finally, just as knowing the cause of fire may enable us to build one when we want one, so knowing the cause of drought may enable us to bring it to an end with prayers and offerings.

This seems a far cry from the scientific search for causes, and in a cultural sense, no doubt it is. But the spirit seems much the same: it embodies the conviction that there is an explanation for everything (the buck always stops, someone is always responsible) and the hope that dedicated inquiry may give us a way of controlling the future. Understanding causal relations gives us power to change our world both directly, through our own actions, and indirectly, through the actions of others. That is because the corresponding causal chains end in a link that is a voluntary action. It is *I* who rub the sticks together, *you* who dump water.

To return to the fire in the forest, the causal agents of certain things that I understand are unknown to me. (I find the fire, but I don't find the person who built it.) Furthermore, there are certain things I now understand (how to build a fire) that I did not understand at an earlier time. Putting such facts together leads naturally (but perhaps not inevitably) to the superstitious postulation of the initiators of causal chains whose effects I experience. And, of course, the obsequious solicitation of more desirable effects.

This same conception of causality as tied up with agency carries over, I suggest, into the philosophy of science and (perhaps) into science itself. It is true that not all causes are such that *we* can manipulate them; we may know the cause of a disease without being able either to cure it or to prevent it. But as in the primitive sketch, to know the cause is to see the *possibility* of manipulation. For me to know that you built the fire is for me to be in a position to ask you to put it out. To know that the gods of the hunt have sent the elk away from our hunting grounds is to be in a position to ask them to bring them back. To know that sympathetic vibrations were the cause of the failure of a bridge is not to know how to avoid them in building the next bridge, but it is the first step toward that knowledge. To know that a certain organism causes the disease that concerns us is to know that *if* we could destroy that organism, *then* we could prevent or cure that disease.

Causes are crucial for constructive engineering; but are they essential for science? In one clear sense, it seems that they are not. Fundamental science rests on axioms or "laws," and those, it seems, neither have nor demand explanation; it is not reasonable to ask for the cause of gravity in Newton's system or for the cause of space in Einstein's. One does not ask why God dislikes sin. But granted that we cannot (or should not) ask for the cause of the stage, we may ask for the cause of the actions of the actors on that stage. And we may be led—or misled—to suppose that indeed there is a cause for every action, either as a matter of psychological necessity or as a matter of metaphysical conviction.

Deterministic Uniformity

The starry heavens above constitute the domain where causality appears to rule supreme. What could be more exactly determined and, furthermore, determined by causes we understand better than the courses of the heavenly bodies? Or to adopt a yet more familiar and comfortable stance, imagine this question being asked before Einstein, when it was generally believed that only modest refinements in the Newtonian system of the world would ever be required. I have already cited LaPlace's answer; should we not strive to emulate LaPlace's supreme intelligence?

It is surely a small step to pass from knowledge of the trajectory of the moon about the Earth (even that, of course, is not quite so simple and elegant as the trajectory of a planet about the sun) to knowledge of the trajectory of a cannonball on a firing range. (Let us leave the exigencies of battle to one side.) Of course, we can predict the course of the moon more accurately than the fall of the shot. But we don't know *exactly* where the moon will be at a certain time, because our prediction is based on imprecise measurements, and we don't know exactly where the cannonball will land, because the flight of the cannonball is subject to minor influences—the influence of the wind, of a variation in the quality and amount of the charge in the cannon, of a variation in the diameter and regularity of the ball and the bore, and so on, as well as imprecision in the measurements of the initial conditions. But if we knew what these influences were—or ideally, if they were nonexistent: the wind does not blow, the powder is measured precisely, and so on—we could predict with precision the fall of the shot. So, one might argue, the two predictions are alike after all, since the orbit of the moon is not known with infinite precision; the measurements on which this knowledge is based admit of error.

Alternatively, one might look at the actual rather than the ideal: in actual fact, we are imperfect in our predictions concerning the moon (though we do improve them as time goes on by taking account of more factors), and we are imperfect in our predictions concerning the flight of the cannonball (though we do improve them as time goes by taking account of more factors). But both phenomena are dealt with as instances of phenomena to which classical dynamics applies. Idealized within classical dynamics, we get close enough predictions in our actual corpus to justify speaking of determination in a constructively local sense (the trajectory of the moon is determined roughly by the masses of the Earth, moon, and sun). Furthermore, in the case of the cannonball, we know what we can vary to change the trajectory—add more powder, alter the direction of the barrel, and so on—and in tune with our initial construal of causality, we know the causes of the trajectory of the cannonball. If we could change the mass of the Earth, we could alter the trajectory of the moon, too; and in the same counterfactual way, we could alter the trajectory of the cannonball by altering the mass of the Earth.

We can give an exact translation of such counterfactuals in our metalinguistic framework. Given our actual metalinguistic corpus **MK***, delete from it all observations of the moon's orbit, the tides, and so on. (Determining exactly what is to be deleted is no small job, but we might as well leave some work for the handy phrase "and so on.") Add some observations (and, of course, it is going to be a matter of some computation to determine what they should be) from which new and larger figures for the mass of the Earth will be derived. In this new set of corpora, the evidential corpus will contain new statements about the mass of the Earth, and therefore also new statements about the orbit of the moon.

Thus we replace what is essentially a counterfactual speculation by an actual conditional that we can (in principle) implement, because it is just (!) a matter of constructing the appropriate syntactical objects. As one might suppose, there is a fair amount that one might say about the logic of such constructions. This approach has been taken by Peter Gardinfors, David Mackinson, and others as a way of explicating counterfactual conditionals and analyzing scientific change.[6]

Now let us consider a third phenomenon: the fall of a dry leaf from a tree in the fall. (Or, if you prefer, the traditional sparrow.) The leaf, like the moon and the cannonball, traverses a certain trajectory. Most people, I assume, would say that it is subject to the same dynamic laws as those solider objects. And yet the idealization does not yield even an approximate trajectory for the leaf. It does no better than an untutored guess: the leaf sooner or later comes to rest on the ground. Nor is there any obvious way to approach knowledge of this trajectory. What measurements might we make that would give us more accurate knowledge of its shape? Nor is there any obvious way to alter the trajectory. What would we change (other than such things as the mass or shape of the leaf, or the direction and velocity of the air currents, which would give us a different problem) that would make the trajectory closer to our heart's desire? Thus, even hypothetically, it is not clear how the manipulative notion of causality can apply to the fall of the leaf. Of course, in the case of God and the sparrow, God (allegedly) does know its fall; that is presumably a matter of omniscient foreknowledge, and not a computational matter or a matter of greater familiarity with the laws of dynamics and kinematics than is available to mere mortals.

Note that we are not considering the possibility that the laws of dynamics as we know and love them do not apply to the falling leaf. Of course those laws apply, on the view being defended here, precisely because they are in a certain sense without content: they specify how our scientific language is constructed and how it is to be used. In just the same way, these laws apply to the cannonball, about which we can

---

[6]Peter Gardenfors, "Belief Revisions and the Ramsey Test for Conditionals," *Philosophical Review* **95**, 1986, 81–93.

only make imperfect predictions; but in the case of the cannonball, we can more accurately evaluate (measure!) those forces required to account for the deviation of its behavior from the more idealized behavior. It is exactly this that we cannot do in the case of the leaf.

This does not mean that we cannot specify the "causes" or the determinants of the trajectory in *general* terms: we can say that the trajectory is determined by the mass distribution of the leaf, the velocity and viscosity of the breezes blowing on it, the dynamics of the fluid flow across its curved surface, and so on. But saying this is just saying no more than that it is such factors, according to our a priori theory of dynamics, that are allowable in explaining and predicting the motion of inanimate objects.

The question is whether it adds content to call these generically characterized factors "causes." I claim that it does not, and that in fact—so far as the metaphysical thesis of universal causality is concerned—it is misleading, since in other cases where we specify causes, we can *get at them*: we can measure them, often alter them, mold them so as to cause (as agents) the effects we desire; at the very least, we can alter them hypothetically by considering corpora syntactically similar to our actual corpora.

It may be enlightening to compare the fall of the leaf with the flight of the sparrow. We certainly suppose that the flight of the sparrow, like the fall of the leaf, is subject to the constraints of aerodynamics. But we do not suppose that more accurate measurements (barring some preposterously hypothetical neurological measurements!) would yield the bird's trajectory.

In fact, does classical dynamics require that the fall of the leaf be caused at all? It seems not. It may require that it be determined: that is, construed as a localized event, its actual outcome is determined by local initial and boundary conditions expressed in the vocabulary proper to dynamics. But even here it is only the idealized counterpart of the actual leaf event that is determined by the laws of dynamics. And this idealized counterpart is completely general; it does not involved the shape, wind currents, mass, and other factors applicable to our *particular* leaf. According to one conception of causality, it is only the behavior of the *particular* leaf that can be said to be caused (by particular occurrences of wind, etc.), and we have no window into this at all. In the general case of a leaf on a breezy day, we not only have no approximation to an ideal with which our dynamic theory could deal, we have no way of obtaining one.

It may be that we should still hold the belief that there exist values (forever inaccessible to us) of the dynamic variables that govern the leaf's trajectory, and that there is a causal connection between the values of these variables and the trajectory of the leaf. That is, perhaps we are warranted in thinking that the only thing that stands between us and the clear-sighted intelligence referred to by LaPlace is the fact that we do not know the positions of the particles and the forces acting on them. Is the problem we face a problem of measurement? In the following section, we shall explore the question of whether one can measure with arbitrary precision, and so achieve (in principle) arbitrarily precise predictions. This is the question of whether universal causation in classical physics (we leave quantum mechanics to one side) is a limiting notion—one that applies, or that we should believe applies, to physics in the limit.

Measurement and Causality

Let us look at the thesis that, if only we knew enough, we could make accurate predictions to the extent that we could make accurate measurements, control the course of events where control could result from measured modifications of physical quantities, and so on. The first thing to note is that this makes the theory of measurement part of the thesis of deterministic uniformity.

We must suppose that the error we make on a given measurement has its causes, too. But we generally have no handle on those causes. We arrive at a theory of errors of measurement of a certain sort, in the final analysis, empirically: by looking at the distribution of results of measuring the "same" quantity. We have already seen how this procedure is possible. But we have also seen how it depends on the conventions we make or the laws we accept (the additivity of length, for example) in the context of measurement.

If statistical sampling of a class of measurements is the source of our knowledge of the distribution of measurements, then we must understand that our knowledge of that distribution is itself (a) only probable, reflecting the uncertainty of statistical inference, (b) approximate, and (c) quite unlikely to give us any ultimate insight into the causes of our errors of measurement.

The first two facts are related: the rougher the approximation we are satisfied with, the higher the probability of the result of the inference that yields it (on given data). But whatever the evidence on which our theory of the distribution of errors is based, the statistical knowledge representing it is both uncertain (present in a corpus of level p but in no higher corpus) and approximate (what we have a right to believe is that the distribution of error is one of a certain family F of distributions). Specifying the family F is hard to do with any precision, but one way that is not implausible in real life is to say that errors are distributed not too differently from some normal distribution with a small mean and a variance that lies in a certain interval.

The third fact reflects the fact that the underlying causes of errors of measurement are irrelevant to the statistical treatment of those errors. That is, if we can identify the cause of a certain part of the error of a measurement procedure, we can eliminate it or take it into account quantitatively. Then it is the *remainder* of the error that we need statistics for.

What might be meant by the claim that the errors of our measurements can (from the point of view of classical physics) be arbitrarily reduced? What is being assumed? It is a handy convenience, rather, to talk of accepting with high probability an approximate distribution of error, to talk of the distribution of errors of a certain instrument as being distributed in a certain way—for example to be normal with a mean of 0 and a variance of s^2. If this were the case—if the distribution of errors of measurement of this instrument were normal, with a mean of 0 and a variance of s^2—then the distribution of the error of the average n measurements of the same quantity would be normal, with mean 0 and variance s^2/n, provided that the errors are independent of one another. We may therefore achieve arbitrary precision by replicating our measurements, subject to these assumptions.

What changes when we replace these assumptions with more realistic ones? We do not actually know what the variance is s^2, but with a given degree of confidence we may know that it is less than a certain amount, given a certain body of statistical evidence concerning it. But this statistical evidence assumes a certain background of physical knowledge, and as we replicate our measurements of the same quantity, we will be getting new evidence concerning the variance of the error distribution.

We do not know that the mean of the error distribution is 0, but only that it is close to 0. (Obviously, to know that it differed significantly from 0 would lead us to introduce a correction factor; I assume that that sort of thing has been done.) But to know that it is in an interval of width 2 around 0 imposes a limit on the accuracy of our measurement even with replication. For if the true mean of the error distribution is d, then (assuming that all the other conditions are satisfied) the limit approached by the average of n measurements as n increases is not the true value, but d units removed from the true value.

Thus, to be sure (and the sureness is less than certainty) that the mean of our distribution of errors is less than δ in absolute value is, at best, to be sure that the average error of measurements replicated arbitrarily often is in the interval $d \pm \delta$, not that it is close to 0. Furthermore, while the measurements in that series of replications cast new light on the variance characterizing errors of measurements of that sort, those measurements do not give new information concerning the mean of that distribution of error.

Finally, we have supposed that the errors of measurement on the various replications are independent of one another. Is this a statistical hypothesis that can be rendered no more than probable by the finite evidence we have? Does it mean that the distribution of the error of measurement i, given certain values of the errors of the first $i-1$ measurements, is the same as the distribution of the error of measurement of the first measurement? This sounds like a lot to know; in fact, on the view presented here concerning probability, it is not so much. For a set of n errors to be "independent" is for that set to be, relative to what we know, a random member of the n-dimensional cross-product of the set of all such errors with itself. In a state of ignorance, this is not hard to achieve, particularly since we generally have no way of getting at absolute errors. But a long sequence of measurements or related quantities provides precisely the sort of evidence that *can* give us reason to think that there are interactions among the errors of measurements in the sequence. Thus, in the arbitrarily long run, we are almost certain to discover reasons to reject the "independence" of the errors of measurement, that is, reasons to put our sequence of measurements in some subset of the corresponding cross-product.

This is to be contrasted with a view that construes independence among the measurements statistically. In the first place, we cannot be sure that measurements are independent in this sense. In the second place, if we assume that they are independent, we are assuming that even in the arbitrarily long run, we cannot learn otherwise.

The upshot of these considerations is that even if we take for granted the framework of classical physics, and we ignore the difficulty that errors of measurement should themselves fall within this framework, we have no grounds for supposing that measurement and therefore prediction can be made arbitrarily precise, and therefore no grounds for assuming that deterministic uniformity holds for the specific process that concerns us. An approximation is as close as we can get, even in the long run.

Another way of putting the matter is this. Consider LaPlace's intelligence, and suppose that it knows the positions of all the particles in the universe, and the forces acting on them, only approximately. That is, humanize it to the point where it can only make measurements eternally subject to error. No longer is the whole past and future of the universe present to it. The present is known to a close approximation, let us say; but since the errors of measurement in some cases will augment themselves (just as, in some cases, they wash out, as in statistical mechanics, where, for example, it is only the *average* kinetic energy that concerns us), there will be, proceeding toward the past as well as toward the future, an expanding cone of approximation. If we are a little bit off in our assessment of the position of the moon, we will be only a little more off in our prediction of its position tomorrow, but very far off indeed in our prediction of its position 1000 years from now.[7]

Within limits, to an approximation, in special cases (such as provided by a satellite or a cannonball), deterministic uniformity not only comes close to holding but is extremely useful. It gives us a handle on both prediction and control. We know what to do to the elevation of the muzzle in order to correct for an error in our aim; we know how to predict the period of an orbit. But the suggestion that we have grounds for universalizing these facts, even without considering the probabilistic character of quantum mechanics, is entirely groundless.

Perhaps few philosophers of science take universal deterministic causation to be either true or important anymore, though many seem to take it as a sort of ideal paradigm. Quantum mechanics has suggested to many that if there is to be some sort of universal causal principle, it must be construed statistically. And, if it is so construed, we can see an explanation for the appeal of universal deterministic causality: it is that if we look at macroscopic phenomena or large numbers of events, we get probabilities very close to 0 or to 1. And this is what misleads us into thinking that the law of universal deterministic causality reigns supreme. But is the statistical version true? That is what we will consider in the next chapter.

---

[7]Except accidentally, like Lewis Carroll's watch that is accurate twice a day.

12

Statistical Causality

Causal Laws

One response to the fact that it is very difficult to find persuasive grounds for believing in universal deterministic uniformity (much less universal causality) has been to suggest that the reign of causality is indeed universal—all events are caused—but that (particularly in the light of quantum mechanics) many, if not all, causal laws are probabilistic or statistical in character. Thus a causal law does not spell out what will be the effect of a given cause in a given case; it just gives a probability of a given effect when the cause is given. This seems appropriate not only in the rarified air of quantum mechanics but especially in the swampy territory of human activity: in economics, psychology, epidemiology, and the like.

One way of construing statistical relationships as causal is by means of the law of large numbers. If under circumstances C event E occurs, not all the time, but $100\,p\%$ of the time, it may seem that the circumstances C do not determine whether or not the event E occurs, and so cannot be said to cause it. (This claim strikes me as absolutely correct.) But, it is suggested, if we consider a large number of repetitions of C, it is almost certain that almost $100\,p\%$ of those repetitions will yield E. This is, so to speak, no more than a hop, skip, and a jump from the case of universal succession (cause being *always* followed by effect). So what we might say is that circumstances C exhibit a causal *tendency* of degree p) to produce E.

There are two main ways in which this notion of statistical causality has been explicated. The first, due largely to Carl Hempel,[1] requires

[1] Hempel, Carl G.: "Deductive-Nomological vs. Statistical Explanation," in H. Feigl (ed.), *Minnesota Studies in the Philosophy of Science*, Vol. III, University of Minnesota Press, Minneapolis, 1961, pp. 98–169.

that p be large: that the cause render the effect highly probable. The second, due largely to Patrick Suppes[2] and Wesley Salmon,[3] requires not that p be large, but that it be larger than it would have been in the absence of C: what is required is that the cause render the effect more probable than it would have been otherwise.[4] In both approaches, there are complications to be taken account of.

Leaving metaphysics to one side, there are three major philosophical contexts in which we often think we have to get clear about causal relations. One of them is the context of explanation: to explain an event is, one might think, to uncover its causes. This is the slant from which Hempel approached the question of statistical causality. If we have good explanations that depend on universal uniformities (causal laws?), then we should have moderately good explanations that depend on near uniformities—statistical laws mentioning parameters that lie near 0 or 1.

Another context is that of decision theory. In decision theory we are concerned to maximize the value—usually the expected value in the sense of probability theory—of our acts. To do this requires that we take account of the causal consequences of the acts open to us. It is here that the contrast between evidential and the other interpretations of statistical causality has been most clear: my choice of an act may be *evidence* for a bad state of affairs without *causing* that state of affairs to be bad.[5]

There is a third context in which the importance of causality is so pervasive and central as to be almost unnoticeable, and that is engineering. In order to achieve a certain goal, we devise a system of causally interacting entities: when the operator pushes the button, that causes a circuit to be closed, which allows electricity to charge the grid under the rabbit, which causes the rabbit to jump, which causes the

---

[2] Patrick Suppes, *A Probabilistic Theory of Causality*, North-Holland, Amsterdam, 1970.

[3] Wesley Salmon, *The Foundations of Scientific Inference*, University of Pittsburgh Press, Pittsburgh, 1967.

[4] Suppes' comment on p. 10 (op. cit.) "that one event is the cause of another if the appearance of the first event is followed with a high probability by the appearance of the second" is abandoned without comment two pages later.

[5] Alan Gibbard and William Harper, "Counterfactuals and Two Kinds of Expected Utility," in C. Hooker, J. Leach, E. MacClennen (eds.), *Foundations and Applications of Decision Theory*, Reidel, Dordrecht, 1978, pp. 125–162.

seesaw on which the rabbit is sitting to rise, which causes . . . This is quite plain in Rube Goldberg's cartoons and in the case of deterministic causality. But it is also discernible in the case of statistical causality.

Sometimes objects are designed to have a random distribution of effects: fragmentation grenades or, less gruesomely, roulette wheels. More often, the stochastic character of the device is simply unavoidable: there is random slippage in the linkage between the beginning and the end; the sensing devices do not perform with perfect reliability or with perfect precision, and so on. We may or may not be moved to take account of this; it depends on what we are designing for what end. We will return to engineering considerations in a later chapter.

Of course, there are caveats. There are events that always co-occur that are not causally connected. Some are just coincidences (it just so happens that all the coins in my pocket are copper), and some are reflections of a common cause (when the kitchen clock shows 12 o'clock, the hall clock strikes 12 times). And misleading correlations can occur in statistics, too. The most famous is a (possibly apocryphal) correlation between the rainfall in Scotland and the number of missionaries in New Guinea. But, in general, it is alleged that causality is the same sort of thing in the statistical realm as it is in the deterministic realm.

One way of characterizing probabilistic causality is to say that "explanations of particular facts or events by means of statistical-probabilistic laws [are such that] the explanans confers upon the explanandum a more or less high degree of inductive support or of logical (inductive) probability."[6] Another way of construing probabilistic causality is to suppose that if there are two populations differing only in that the causal factor is present in 100% of the cases in the first population and in 0% of the cases in the second population, the effect will occur more often in the first population than in the second.[7] A third way of construing probabilistic causality is to require that the probability of the effect, given the cause, be greater than the probability of the effect alone (or given the denial of the cause) and "that there

---

[6]Carl Hempel, *Aspects of Scientific Explanation*, The Free Press, New York, 1963, p. 385.
[7]Ronald Giere, *Understanding Scientific Reasoning*, Holt, Rinehart, and Winston, New York, 1979, p. 180.

be no third event that we can use to factor out the probability relationship between the first and second events."[8]

In the following sections, we shall explore these ways of construing probabilistic causality. But first, we should observe that there is a radical departure from the more traditional notion of causality entailed by going from deterministic to statistical causality. On the traditional view, one event causes another when the first exhibits some *power* that produces the second event. On any of the views just alluded to, it is not individual events that stand in the causal relation, but classes of events. Giere is quite explicit about this: "the best way to understand such [causal] claims is as not referring to individuals directly but to a whole population of individuals."[9] Suppes writes, "When we put the matter this way in terms of relative frequency, we are of course speaking in terms of kinds of events."[10] (But then he follows this statement with the utterly cryptic remark, "We can by the usual process reduce this discussion of kinds of events to an analysis of particular events.")

One does not have to be much of a nominalist to have severe doubts about a class of events causing a particular kind of event, or a relative frequency in a class of events causing a particular kind of event, or one class of events causing a certain relative frequency in another class of events.

Some writers have suggested that what is really caused when we speak of C as being a statistical or probabilistic cause of E is a *disposition* to produce E. Thus one might say that tossing the coin produces deterministically, and every single time, a disposition (of strength $1/2$) for a coin to yield heads.[11] What is caused, then, is universal and not statistical after all.

But what is caused is no longer the sort of thing we can observe; you can't observe a disposition directly, though, of course, you can observe its statistical manifestation. There is nothing wrong with this

---

[8]Suppes, op. cit., p. 10.

[9]Giere, op. cit., p. 176.

[10]Suppes, op. cit., p. 45.

[11]James Fetzer and Ronald Giere have both discussed chances as dispositions. See in particular R. Giere, "Objective Single Case Probabilities and the Foundations of Statistics," in Patrick Suppes et al. (eds.), *Proceedings of the Fourth International Congress of Logic, Methodology and Philosophy of Science, Logic IV*, North-Holland, Amsterdam, 1973, pp. 467-484; and J. H. Fetzer, "Statistical Probabilities: Single Case Propensities vs. Long Run Frequencies," in W. Leinfeller and E. Kohler (eds.), *Developments in the Methodology of Social Science*, Reidel, Dordrecht, 1974, pp. 337-347.

kind of theoretical interpretation of probabilistic laws — it seems quite likely that the best interpretation of quantum mechanical laws would be along such lines — but it seems to involve far more superstructure than is warranted when it comes to coins and causes of death in various populations.

Maximal Specificity

Carl Hempel's general scheme for explanation may be represented as follows:

$$\frac{(x)(Cx \rightarrow Ex)}{Ca}$$
$$\overline{Ea}$$

The first premise is construed as a law, suggestively written to hint at a causal relation between C and E; the arrow need not be construed as the so-called material conditional. The second premise is the claim that a satisfies the predicate C; and the explanation as a whole consists of the derivation of the conclusion that a satisfies the predicate representing the effect.

There is a long history of debate over the adequacy of this notion of explanation. We need not enter into that debate — the debate over whether or not this is really *explanation* — for what concerns us mainly is what happens to this schema in the attempt to generalize it to take account of statistical explanation. The schema for statistical explanation offered by Hempel looks like this:

$$\frac{\%(E,C)=p}{Ca} \quad p$$
$$\overline{Ea}$$

The interpretation is this: the frequency of E's among C's is p, where p is reasonably close to 1 or "high"; a is one of those C's. So we conclude with confidence p that a will also be an E. Note that the conclusion is categorical: Ea. It is not of the form "probably Ea."

We may also consider a schema in which the conclusion is not "Ea" but "the probability of Ea is p." Since Hempel is looking for an explanation of "Ea" and not of "the probability of Ea is p," the former schema is the one that primarily concerns us. But in either case, something more is required than has been reflected in the schema.

In the case of inductive or statistical explanation, in contrast to the case of deductive or nomological explanation, we require something like a principle of total evidence. If we add to our first schema an additional premise, "Da," it does not undermine the explanation of Ea, because it does not jeopardize the deduction of "Ea." (This is not quite true, and in fact constitutes one of the objections to the argument notion of explanation. If we are offered an explanation of A's death in terms of her fall from a high bridge — suitably beefed up to yield the deduction that "A is dead" — that explanation *is* undermined by the added premises that she ingested cyanide before she fell, and so died before hitting the ground. But it is true that adding premises does not compromise the deduction.) On the other hand, it is obvious that the addition of a premise can undermine a statistical argument.

For example, we may explain the fact that Jack is friendly in terms of the fact that 90% of longshoremen are friendly and the fact that Jack is a longshoreman. But this explanation no longer works when we discover that Jack is a longshoreman who has just lost his job for fighting and that 90% of those longshoremen who have lost their jobs for fighting are unfriendly.

Hempel's response to this kind of difficulty is to impose a *rule of maximal specificity*. Without going into details, we can capture the gist of the idea as follows: The explanation of "Ea" on this basis of "Ca" works only if for every predicate "C'" such that "$(x)\ (C'x \rightarrow Ca)$," we know that the proportion of C' that are E is also p: "$\%(C',E)=p$." Hempel excludes the case where "$\%(E,C')=p$" is a theorem of probability theory.[12] Other special cases must be excluded as well.[13] Thus it is a condition of an adequate statistical explanation that we take account of everything we know about the subject of the explanation that is relevant to the "effect" E, and that the probability be high.

As it is stated, Hempel's criterion renders statistical explanation hard to come by. In the case of Jack, being Irish may account for his friendliness: 90% of Irish longshoremen who have just lost their jobs are nevertheless friendly. We have saved the explanation. But surely an

---

[12] C. G. Hempel, *Aspects of Scientific Explanation*, The Free Press, New York, 1963, p. 400.

[13] H. E. Kyburg, "More on Maximal Specificity," *Philosophy of Science* 37, 1970, 295–300. This issue has arisen again in the context of inheritance hierarchies and nonmonotonic logic in artificial intelligence. See H. E. Kyburg, "Epistemological Relevance and Statistical Knowledge," in Shachter and Levitt (eds.), *The Fourth Workshop on Uncertainty in Artificial Intelligence*, 1988, pp. 237–244.

appropriate property to consider is that of being over 60, and Hempel's condition would require that we *know* that 90% of the Irish longshoremen who have just lost their jobs for fighting and are over 60 are friendly. But we may simply know nothing about this rather restricted and specialized class. And if P and Q are appropriate predicates, so ought their conjunction to be, which leads to exorbitant demands on our knowledge.

More realistically, we may know something, but perhaps not enough to assign a probability different from .90 to Jack's friendliness. More realistically yet, we may want to take account of knowledge asserting that the frequency of E's among C's lies in a certain interval.

But then we have exactly the problem of the reference class, as we discussed it earlier. And the same controversies. David Papineau[14] asserts that the maximal specificity requirement is equivalent to the requirement that the reference class from which we obtain our probability be homogeneous. Homogeneity is a notion that has also been suggested by Wesley Salmon.[15] Salmon takes C to be a homogeneous reference class for a, provided that there is no property P that a has (that a can be ascertained to have without knowing whether or not a has E) such that the frequency of E among items in the intersection of C and P differs from its frequency in C. Homogeneity is intended by Salmon as an objective notion: C is homogeneous for E or not, whatever we may think or know. But as a *practical* matter, we may accept a reference class as homogeneous when we don't *know* of any way to divide it usefully. This is *epistemic* homogeneity.

Hempel's notion of maximal specificity appears to be an epistemic notion, since Hempel relativizes it to a set **K** of propositions that we can interpret as a body of knowledge. But even in the case of objective homogeneity, we cannot consider all predications or all classes, since there is always a set (predicate) to which a belongs (that a satisfies) in which the frequency of E is 0 or 1: namely, the singleton of a (the predicate "$=a$"), in which the relative frequency of E is different from its frequency in the putative reference class C. Furthermore, this even undermines *epistemic* homogeneity, since we know that a belongs to this set.

If it were only a matter of singleton sets or identity predicates, we could dismiss this problem as an artifact of our logic. But we can

---

[14] David Papineau, "Probabilities and Causes," *Journal of Philosophy* **82**, 1985, 57–74.
[15] Salmon, op. cit.

always find some set satisfying whatever natural constraint is proposed that has an undermining effect. Should we reject singletons? How about the set of all x such that $x=a$ or x belongs to both C and E, which yields a reference class with a relative frequency of E close to 1? Should we demand that reference classes be infinite? We can arrange that. Should we demand that it be an empirical matter whether a belongs to the reference class or not? Consider the set of all x such that x is observed on Tuesday and . . . (where a is observed on Tuesday).[16]

It was exactly considerations of this sort that concerned me in *The Logical Foundations of Statistical Inference*, and *Probability and the Logic of Rational Belief*. In both of those works I proposed that we begin with a language, and that we recursively define a set of terms that are to serve as the potential reference terms of the language. (Recursively: we must have a mechanical procedure for deciding whether or not an arbitrary expression is a potential reference term of the language.) For example, it seems reasonable to demand that the intersection of two reference terms ought to be an acceptable reference term, and we can embody this intuition in our characterization of reference terms. Similarly, many of the troublesome terms arise from the formation of unions; we do not want the union of two good reference classes to be (necessarily) a good reference class. Again, the complements of plausible reference classes do not generally constitute plausible reference classes.

This relativizes the set of potential reference terms to a language, since this set is defined in terms of the primitive predicates of the language. There are difficulties even with the characterization given in *The Logical Foundation of Statistical Inference*, so for our present purposes, it will be best to take the set of potential reference terms to be simply *given* as part of the language. Since we have a procedure (roughly outlined earlier) for choosing between languages, this presents no new problems. Note that, of course, since we have to be able to specify a language recursively, we must be able to specify the reference terms recursively *within* a language. Furthermore, it seems reasonable to suppose that there are some general constraints we could uncover concerning the specification of reference terms. (In fact, we have already uncovered one: the set of potential reference terms cannot be taken to include every well-formed term in the language.)

---

[16]Kyburg, "More on Maximal Specificity."

Given a set of potential reference terms to work with, there is still the question of what counts as a statistically relevant division of a potential reference class. The condition that the division be regarded as statistically relevant unless we *know* that the frequency in each component is the same as that in the class itself seems excessively strong. (But this is exactly Isaac Levi's requirement for direct inference.[17]) If we deal with knowledge of intervals (as I suggested earlier), the situation becomes even more confusing, as I shall show in the next section.

"Screening Off"

It is arguable whether we really require high probability for statistical explanation. As Michael Scriven pointed out in an example that has become a classic,[18] the presence of paresis in a patient is explained by prior infection with syphilis, even though only 3% of syphilitics ever proceed to this tertiary stage of the disease. In the case of statistical treatments of causation, it seems quite clear to many writers that causes operate even when the probabilities involved are quite small. A necessary condition for the existence of causation is, as Suppes puts it, that the probability of the effect, given the cause, is greater than the prior probability of the effect: $P(E|C) > P(E)$.[19] (He also adds the conditions that the cause precede the effect and that the cause have positive prior probability.)

But this condition is not enough, by itself, for it amounts to demanding only a correlation between cause and effect, and as we all know, there are misleading correlations. So we need an additional condition, and the one that is most generally proposed is that there be no property G that *screens off* E from C. (I believe that the phrase "screen off" is due to Brian Skyrms.[20])

Here is the classical example of screening off. Suppose that there is a correlation between smoking S and cancer C. $P(C|S) > P(C)$. Equiva-

---

[17]Isaac Levi, *The Enterprise of Knowledge*, MIT Press, Cambridge, Mass., 1980, contains an extensive and enlightening discussion of direct inference, including a critical comparison of Levi's procedures and mine.

[18]Michael Scriven, "Explanations, Predictions, and Laws," in H. Feigl and G. Maxwell (eds.), Minnesota *Studies in the Philosophy of Science*, Vol. III University of Minnesota Press, Minneapolis, 1961, pp. 170–230.

[19]Suppes, op. cit.

[20]Brian Skyrms, *Causal Necessity*, Yale University Press, New Haven, Conn., 1980.

lent conditions are $P(C|S) > P(C|\sim S)$ and $P(C\&S) > P(C)*P(S)$. This might not establish a causal connection between smoking and cancer if, as Sir Ronald Fisher conjectured,[21] both smoking and cancer were correlated with a certain genetic makeup G: people prone to cancer are also prone to smoke. Formally, if

$$P(C|S\&G) = P(C|G)$$

and

$$P(C|S\&\sim G) = P(C|\sim G),$$

we say that G *screens off* the association between S and C.

(We are leaving out of account the temporal relation, but it should presumably be included. It is not reflected directly in any of these probability relations; so far as they are concerned, we can as well say that dying of cancer is itself the cause of smoking.)

As noted by Papineau,[22] satisfaction of the rule of maximum specificity, or the requirement that the reference class be homogeneous, implies that there is nothing to screen off the alleged causal relation, but the converse implication does not hold. (Either approach, of course, may be construed either epistemically, as relative to a corpus of knowledge **K**, or nonepistemically, as relative to statistical truths about the world.)

To see the implication from homogeneity to no screening off, note that if there is no (acceptable) subclass of C in which the frequency of E differs from that in C, there can be no set G such that G screens off C. To see the failure of the converse implication, note that we may divide C into subclasses $C_1, \ldots C_n$, and as long as there is no screening off in any of these classes, there will be no screening off in C in general, even if the frequency of the effect varies a lot between one subclass and another.

This is a suitable place at which to mention Simpson's paradox, in which a genuine correlation in each of a number of subsets of C can be hidden, or even reversed, in C itself. Let C be divided into two parts, C_1 and C_2. Suppose that G is a cause and E is the corresponding effect. We can have a situation in which

---

[21] R. A. Fisher, *Smoking: The Cancer Controversy*, Oliver and Boyd, London, 1959.
[22] Papineau, op. cit., p. 61.

$$P(E|G\&C_1) > P(E|C_1)$$

and

$$P(E|G\&C_2) > P(E|C_2),$$

so that in each part of C, G encourages E, and at the same time,

$$P(E|G\&C) < P(E|C).$$

This seems counterintuitive. A numerical example will show that it is perfectly possible, though:

C_1	E	G	$G\&E$
1800	90	1000	60
C_2	E	G	$G\&E$
200	60	10	4
C_1	E	G	$G\&E$
2000	150	1010	64

It is easy to verify that $P(E|G\&C_1) > P(E|C_1)$ and $P(E|G\&C_2) > P(E|C_1)$, but that $P(E|G\&C) < P(E|C)$.

Note that the same requirement to worry about what constitutes a legitimate potential reference class arises in the case of statistical association as in the case of high probability. Without constraints on what sorts of classes we can refer to, we can always come up with a screener-off. The effect class E itself is the universal nonepistemic screener-off, but there are any number of ways of defining arbitrary sets that will function as either nonepistemic or epistemic screens between any alleged cause–effect pair. In fact, since the members of the causal class C can (presumably) be individuated, we can assign them to the *arbitrary* class G in such a way that both

$$\%(E, C\&G) = \%(E, G)$$

and

$$\%(E, C\&\sim G) = \%(E, \sim G)$$

are preserved, while $\%(E, G)$ and $\%(E, \sim G)$ diverge as much as we want.

We are thus led once more to constraints on potential reference classes: only certain terms can plausibly be taken to denote potential reference classes.

This is particularly the case if we take the causal relation metaphysically or nonepistemically. For in this sense we are, as the condition is usually stated, simply quantifying over all sets ("There is a no set G such that . . . "). And in general, one can almost always find such a set or, if not, still show that one exists.

But if we construe the condition epistemically—and the same is true of the rule of maximal specificity construed as demanding "merely" epistemic homogeneity—then we must take account of the fact that we do not know any probabilities with real number precision. This fact has a bearing both on the rule of maximal specificity and on the screening-off relation.

The rule of maximal specificity, construed epistemically, requires that our reference class be homogeneous so far as we know. There are two ways to spell this out. We might demand that for every proper reference term denoting a subset of our reference class, we know that the frequency of the effect is the same as it is in our original reference class. Or we might only demand that for every such subset, the frequency *not* be known to differ. And then we must spell out what it means for a frequency to "differ" when we are talking about intervals.

For example, suppose that the frequency of E in C is known to be between .85 and .95. On the first interpretation of the homogeneity condition, we would require for every legitimate reference class C' known to be included in C that we know that the frequency of E in C' is between .85 and .95.

Presumably, either stronger or weaker knowledge would undermine the requirement of homogeneity. But if there is a reference class C' included in C in which the frequency of E is known to lie between .89 and .91, then presumably C' should be the reference class for our explanation. If there is a reference class C'' included in C in which the frequency of E is known only to lie between .80 and .99, then on this approach C'' should be our reference class.[23]

This requirement seems too strong, though it has been accepted by Levi[24] as a principle for the choice of a reference class. If singletons can qualify as reference classes, then, since our knowledge about them

---

[23] I say "frequency," but nothing hinges on construing our statistical knowledge as knowledge of frequencies here. The same things may be said for chances, with the exception that chances do not depend, as frequencies do, on the cardinalities of reference sets. A singleton reference set can only yield a frequency of 0 or 1, but it may yield any interval of chances.

[24] Levi, op. cit.

is generally represented by the whole interval [0,1], all explanations and inferences are undermined. The weaker requirement that only in the case of *more precise* knowledge is our explanation undermined seems more appropriate.

Suppose that we know that the frequency in C'' is between .80 and .99; this does not conflict with the interval characteristic of C. Clearly, a frequency in the interval [.25,.35] does conflict with our knowledge about C. How about the interval [.75,.86]? How about more precise knowledge — the interval [.89,.91]? The considerations we are led to are exactly the considerations we were led to in devising principles for the choice of a reference class earlier.

Causality as Crutch

There is one other (slightly embarrassing) ground for regarding science as the search for causes, either statistical or universal. It is the idea, not altogether unlike the idea that people will behave themselves only if they believe in eternal damnation, that only if scientists believe that every event has a cause will they pursue their knowledge of the world. It may be that many scientists have this belief, just as it may be that many citizens believe in hellfire and damnation. Nevertheless, the efficacy of the belief in promoting social welfare hardly counts as *rational* grounds for holding it in either case. And in either case, whether or not the belief *is* beneficial is more than a little questionable.

In the preceding chapter, we saw that belief in universal causation is unwarranted by anything we have reason to believe. To accept it, even if we could give it a precise and useful meaning, would not contribute to our corpus of predictive observational, practical certainties. There are also grounds for believing that *even if* there is a way of giving other than hypothetical sense to the "determinism" of the fall of a leaf, there would *still* be no gain in predictive observational content. These grounds, however, concern computability, which we have been leaving to one side.

It is relevant to note that there are relatively simple deterministic programs ("chaotic" programs) whose outcome is just as efficiently predicted by running the program as by taking advantage of its deterministic character. In such a case, even though the predictive truth is entailed by the known circumstances or boundary conditions, and therefore exists in our corpus of practical certainties, to *find* it there requires as much time and effort — as many computational steps — as waiting to see.

In any event, it does us no good to assume universal deterministic uniformity. Is the statistical version more useful? Not obviously. In fact, it is not clear that, on the statistical relevance view, the claim of universal statistical uniformity *has* content.

The two standard approaches to probabilistic causality — the statistical relevance model and what Papineau in "Probabilities and Causes"[25] calls the "standard" model — do not add anything useful to decision theory over and above what is offered by conceptions of manipulability. Causality seems relevant only if, by *doing X*, I can influence the probability of *Y*. Do causal stories add anything to understanding beyond this? When I claim a causal relation between *C* and *E*, it seems that I mean no more than that if I *could* influence (bring about, prevent) *C*, I could *thereby* influence (bring about, prevent) *E*, at least in a probabilistic sense. Manipulation, perhaps hypothetical manipulation, seems to be the key.

But if this is our view of causality, then the universalization of the view is indeed superstition: it is the view that with regard to any state of affairs *X* (rain or drought) there is some entity that can, by appropriate manipulation, change the probability of that state of affairs.

Newcomb's problem[26] will serve to illustrate what I mean. Let there be two boxes, one opaque and one transparent. Through the top of the transparent box, we can see that there is a thousand dollars in the box. We are free to take the contents of the opaque box only or the contents of both boxes. But the catch is that we are reliably informed (How? Never mind!) that an infallible predictor has placed a million dollars in the opaque box if and only if he has predicted that we will take only one box — that is, the opaque one.

The puzzle is that if we look at the situation from an evidential point of view, our taking both boxes provides (conclusive) evidence that the opaque box is empty, while our taking only the single opaque box provides (conclusive) evidence that there is a million dollars in it.

But if we think about the "causal structure" of the problem, the million dollars either is (if the demon predicted that we would take only that box) or is not (otherwise) in the opaque box. In the ordinary causal world, there is nothing the demon can do about it if we *now*

---

[25] Papineau, op. cit.
[26] First publicized in R. Nozick, "Newcomb's Problem and Two Principles of Choice," in N. Rescher (ed.), *Essays in Honor of Carl G. Hempel*, Reidel, Dordrecht, 1969, pp. 114–146.

decide to cross him up and take both boxes. And, of course, if the demon predicted that we would take both boxes we are not only no worse off, we are a thousand dollars better off if we take both boxes.

This puzzle has generated an extensive philosophical literature, as well as a certain amount of interest on the part of decision theorists in economics. It is beginning to generate interest in the field of artificial intelligence. The literature is primarily a contest between "two-boxers" and "one-boxers"—that is, between people whose intuitions fall on opposite sides in this case. There is serious controversy.

From the present point of view, despite the skepticism concerning causality we have expressed, we come down firmly on the side of the two-boxers. It is a question of manipulation. I cannot manipulate the money into the box if it is not there, or out of it if it is. The demon is no longer under my control, and the boxes are no longer under his control—given our usual well-justified view of the world, and of what can and what can't be manipulated in it.

We do talk about causality in connection with things we cannot manipulate; we say that the moon causes the tides. But this is not so difficult to explain. Consider a corpus of knowledge just like the one we have, except that it contains no sentences concerning the tides. From that corpus, delete everything that pertains to the existence of the moon. This amounts to: pretend that you could abolish the moon. Then there would be only sun tides. The moon causes the tides in the sense that manipulating the moon out of existence would abolish the tides or profoundly change their character.

There is thus a deeply pragmatic aspect to our talk of causality. Manipulation—and degrees of accessibility to manipulation—provides the key to understanding causal claims. While deterministic conceptions of causality appear to be unwarranted—even leaving aside questions pertaining to the quantum mechanical world of microphenomena—statistical conceptions of causality appear to be otiose. Neither the mysterious power of one event to produce another, nor the even more mysterious power of a class of events to have a statistical effect on another class of events, seems to be relevant to our understanding of the world except insofar as it allows us, actually or hypothetically, to alter the course of the world.

Causal relations are important in two contexts: the context of decision theory and the context of engineering design. Both of these concerns will be the subject matter of subsequent chapters.

13
Dispositions and Modalities

Connectives?

Despite our negative conclusions concerning deterministic uniformity or universal causality in the last chapters, there is no denying that less than universal versions of causal principles play a large part in our lives. We expect nature to repeat itself: if sugar dissolves in my coffee today, it will dissolve tomorrow. We also expect to bend nature to our wills: I can make my crops grow better by adding fertilizer and lime to my soil. But, of course, we don't expect all kinds of things to repeat or everything to be under control. The primitive incantation that was followed immediately by rain last summer may not be followed immediately by rain this summer. I have yet to learn how to control the outcome of a spin of the roulette wheel.

Many philosophers, and not a few nonphilosophers, have seen in the relations and connections among events something that should be reflected in relations and connections among statements concerning events. Consider the fact that a spoon of sugar, placed in a cup of hot coffee and stirred, will dissolve. Let S be the statement that a spoon of sugar is placed in a cup of hot coffee and stirred, and let D be the statement that it dissolves. There are contextual matters to be taken account of: of course, we mean cane sugar, and not sugar of lead. Naturally, we are supposing that this is the first spoon of sugar to be placed in this particular cup of coffee; we exclude the possibility that the coffee is already saturated with sugar. We assume that the sugar is free to dissolve (e.g., that the sugar crystals are not coated with a thin layer of waterproof plastic). And there are many other things we assume.

Taking all this context for granted, how do we express this fact? One natural way to express it, in ordinary language, is to say that if the sugar is placed in the cup, it will dissolve: if S then D. Alternatively, one might express it by saying that the fact that the sugar is put in the cup *implies* that it will dissolve. All this is relatively casual talk. What happens when we want to express the fact more precisely?

The first thing that everybody will agree to is that the so-called material conditional (i.e., the truth functional conditional, often rendered formally by "$\rightarrow$") does not capture the connection we have in mind. "$S \rightarrow D$" is true not only when S and D are true but also when S is false. But when S is false, "$S \rightarrow \sim D$" is just as true as "$S \rightarrow D$."

I conjecture that no one would even think of trying to express a connection between facts by a sentential connective but for Whitehead and Russell's monumental mistake in speaking of "material implication." Having discovered that the truth-functional connective "$\rightarrow$" answered many of the needs for a conditional (if . . . then −) in mathematical argument, and being swayed by the fact that "if . . . then −" is often used in mathematical argument to express an implication, they devised this unnatural name for their sentential connective. In itself this would be an innocent confusion, but for the fact that philosophers ever since have been led to search for other (nonmaterial? immaterial?) connectives to express real implications of the sort we are now concerned with.

Somehow we wish to express the fact that the sugar *must* dissolve when we stir it into our coffee. Genuine implication won't do, since that is a relation between sentences or sets of sentences. ("The theorems are exactly the sentences that are implied by the axioms.") So we seek some way of formalizing "if . . . then −" that is weaker than the strong relation of logical implication—the relation that holds between "4 is even and 4 is greater than 2" and "4 is not prime"—and the mere truth-functional conditional "$\rightarrow$". Never mind that looking for something intermediate between a relation and an operator is rather like looking for something intermediate between C sharp and blue.

The search is tied to the development of modal logic. A suitable connective for expressing causal connections is merely one among a number of related desiderata sought by many philosophers. If we had the right conception of *possibility*, for example, we could explicate the relation between S and D by saying that "S and not-D" was not *possible*. Or *necessity*, as in causal necessity: "Necessarily, $S \rightarrow D$." All of these notions are related. Unfortunately, as witnessed by the large number of distinct systems of modal logic, there are a great many distinct intuitions even concerning the properties that these notions should have.

A similar problem, one that has led to a similar proliferation of formal systems, is the problem of counterfactual conditionals. If we focus on the fact that both "$S \rightarrow D$" and "$S \rightarrow \sim D$" are true when S is false, we may be led to look for a connective that lacks this property: a counterfactual conditional that allows at most one of "if S then D" and "if S then $\sim D$" to be true. Or, in English, one that could be used to express the claim: "If the sugar *were* to be put in the coffee (though in fact I like my coffee unsweetened), then it *would* dissolve."

A modern and elegant way to approach such questions (first proposed by Saul Kripke[1]) is through the semantics provided by a consideration of possible worlds. The idea (as elaborated by David Lewis[2] and Robert Stalnaker,[3] among others) is that "if S were, then would-be D" is to be true just in case, in the world nearest to our actual world (or in any world near enough), in which S is true, D is also true. Once we have the truth conditions for these statements, we can find the principles of inference—the logic—of arguments that preserve truth thus defined.

There is one difficulty about this approach. I *know* that if I were to put sugar in my coffee and stir it, it would dissolve. There are two ways I could know this. I could know which world in which sugar was added to my coffee was closest to the actual world, and I could examine that world to see if, in actual contingent fact, the sugar dissolves there. Of course, there are many statements true in that world that are also true in the actual world—for example, the statement that if I were to bet on horse n tomorrow at Hialeah, I would get rich. So if I can confirm the counterfactual conditional by examining a nearby nonactual world, I can also use that nearby nonfactual world to get rich.

Alternatively, the basis of my knowledge of the counterfactual conditional may not be contingent knowledge about a nearby world, but knowledge that whichever one the possible world may be that is closest to our actual world and makes the antecedent of the conditional in question true, it will have to be such—it *must* be such—that the

---

[1] Saul Kripke, "A Completeness Theorem in Modal Logic," *Journal of Symbolic Logic* 24, 1959, 1-14.
[2] David K. Lewis, *Counterfactuals*, Harvard University Press, Cambridge, Mass., 1973.
[3] Robert C. Stalnaker, "Possible Worlds," *Nous* 10, 1976, 65-75.

consequent is true. But then it is not clear that the talk of possible worlds has been enlightening.

The idea that has been pursued seems to have been roughly this: we know that events are causally connected; propositions—or statements—often express the occurrence of events; therefore propositions corresponding to causally connected events should be connected; therefore there should be a connective corresponding to causal connection.

Put thus baldly, the whole idea seems like a poor pun. Didn't Hume show that "necessity" was reflected in the relations of ideas and not in the relations of matters of fact? But there is no doubt in our minds that the sugar *must* dissolve, and no obvious relation between the sentences S and D. D is clearly not derivable from S. But what other kind of "necessity" might hold between sentences?

What I suggest is that there *is* a relation of derivability that captures the relevant sense of "necessity." It is derivability in a corpus of knowledge. Roughly, the idea is that if D (and, of course, everything that implies D) is purged from our corpus of practical certainties, and S is added to the evidential corpus, then D will reappear in the corpus of corporal certainties. This procedure has the advantage that it gives us a place to store the unstated conditions that must obtain if the truth of even the conditional $S \rightarrow D$ is to be upheld: no tricky plastic-coated sugar, no saturated solutions. We don't demand that the *denial* of these possibilities be in the corpus in question, but only that the possibilities themselves *not* be included. This is what is required for (nonmonotonic) inference in artificial intelligence, and it is what is required here.

This renders the "derivability" nonmonotonic, since now a statement to the effect that the sugar grains are plastic coated can be *consistently* added to our corpus and preclude the derivation of D. We glossed over this in the last chapter, but there as well as here, the relevant sense of "derivation" is a probabilistic/epistemic argument from an evidential corpus of premises to a conclusion in a corpus of practical certainties. As we noted earlier, the inclusion of a statement in the corpus of practical certainties is justified by its probability relative to the corresponding evidential corpus. Such probabilistic inference is notoriously nonmonotonic.

It is a nontrivial matter, however, to accomplish the purging of D, since there are any number of ways to change a corpus **K** so that D no longer appears in it. Later in this chapter, we will try to say something useful about this problem. But first, we will look at the probabilistic counterpart.

Probabilities

Let us begin by considering uniformities and counterfactual claims that do not (or do not obviously) depend on universal dispositions or causal laws. I have in mind such statements as "If this die were to be thrown 10 times, it would not land 1 each time" and "If Pete were to go to the race track, he would lose his shirt."

It might be that such counterfactual inferences, being more insecure than those that involve universal, nomological, causal connections, would be more difficult to analyze. The opposite is the case, since there are often fewer difficulties in the way of redesigning our evidential corpus when the connections that concern us are merely statistical.

Consider the example of the die. The die has no doubt been thrown 10 times — or hundreds — but ordinarily (as the claim was stated above), what is *meant* is that if the die were to be thrown 10 *more* times, it would not result in a one each time. Ordinarily, nothing in our evidential corpus stands in the way of adding "The time is now t, and the die is to be rolled exactly 10 times after t." Since we have specified nothing about the die to suggest that it is not fair, we suppose that relative to our evidential corpus, this (hypothetical) set of 10 rolls is a random member of the set of 10-member subsets of the set of rolls-of-a-die in general, with respect to yielding 10 ones, and in our ordinary corpus we know that the frequency of 10 ones among such objects is *about* $(1/6)^{10}$.

Relative to our evidential corpus, then, the probability is about $1-(1/6)^{10}$ that the 10 rolls in question will not all yield ones. But this is a large enough number to warrant including the assertion that the next 10 rolls will not yield 10 ones in our corpus of practical certainties. Or: It is practically certain that the next 10 rolls would not yield 10 ones.

Suppose that we know the die to be loaded? Well, of that known-to-be-loaded die, we might not say that its hypothetical next 10 tosses were a random member of the class in question. (But if it were loaded in favor of two or five, that would only decrease the frequency of ones.)

Suppose that we know that the die will never be tossed after t? See this die? I hurl it into the furnace. *Now*, friend, why should I agree that if it *were* to have been tossed 10 times, it would not have landed 10 ones? To answer this question, one may construct the evidential corpus that is just like yours, except for not containing any statement that implies that the die was tossed into the fire at t or was otherwise destroyed at t. In this corpus we may suppose that the die exists after t and that it is tossed 10 times after t. Once we have this corpus, the previous argument goes through. (Alternatively, the sequence may be regarded as a member of sequences of dice in general.)

Standard, uncontroversial, probabilistic projections do not in general require any modification of the corpus of evidential certainties beyond their augmentation by a sentence describing the object whose property is to be projected. Note that there is nothing counterfactual about the evidential corpus in question: it is an abstract, timeless, syntactical object. The claim is not that if that *were* your evidential corpus, then the assertion that the die will not land heads 10 times *would* be in your (rational) corpus of practical certainties — though that is true. It is that the sentence in question appears in the practical corpus corresponding to an evidential corpus described in a certain way. It is a syntactical claim, though the objects it concerns (our corpus of evidential certainties, our corpus of practical certainties) are implicitly assumed to represent our best guesses about the world.

How about the claim that if Pete went to the race track, he would lose all his cash? Assuming that our evidential corpus contains sentences asserting that Pete did not go to the race track, that Pete was in Ottawa at the time, and so on, we must purge our corpus of these sentences. Exactly how far should we go? For example, if we have "Pete does not go to the race track on date d" in our evidential corpus, we do automatically have such sentences as "If Pete goes to the race track on date d, then Pete loses all his money," as well as "If Pete goes to the race track on date d, then Pete wins a bundle" and "If Pete goes to the race track on date d, then Pete comes home with a horse."

These sentences do not (singly) imply that Pete does not go to the race track (though "If Pete goes to the race track on date d, then $1=0$" does), so if we simply purge "Pete does not go to the race track on date d" from the evidential corpus, we have an awkward collection of conditionals left in that corpus. In fact, we may well also have, in the practical corpus, the statement "Pete does not go to the race track." Some conditionals we do want to hang on to: "If Pete goes to the race track, he takes a bus" is one we might want as an instance of the generalization that everybody (or almost everybody) who goes to the race track takes a bus. This kind of problem could have been raised in connection with the counterfactually rolled die, but we chose not to raise it then for reasons of simplicity.

How could we afford not to? The answer is that the change to be rung in our evidential corpus is completely obvious in that case. We get our new corpus by adding a statement to the corpus we start with: "The die is rolled 10 times after t." It is hardly possible to imagine a simpler change. But in the case of Pete, it may not be quite so simple to make the appropriate change. For example, we might know that Pete's sister is getting married on date d and that Pete would never fail to appear at his sister's wedding. So how do we change the corpus that contains all these things?

This is a classical problem in the analysis of counterfactuals, and I have no profoundly new recipe. The current fashion is to make the counterfactual conditional with antecedent A and consequent C true (here and now, in this world) just in case C is true in the nearest possible world (or all the closest possible worlds) in which A is true. It is difficult to judge which possible worlds are near to which, and so there is a certain amount of vagueness associated with the analysis. It is claimed (e.g., by David Lewis[4]) that this is appropriate, since the notion being analyzed is vague. (But then what do we gain?)

There is a certain amount of vagueness in a counterfactual antecedent—that is, it is hard to know exactly what changes in our present corpus are intended. If we think of truth conditions semantically, in terms of possible worlds, it is not clear that we can always reduce this vagueness. But if we think of counterfactuals as metalinguistic and relativized to corpora of sentences, then we can (more easily, at least) specify as carefully as we wish what (syntactical) changes are to be made in our evidential corpus. And then it is a question of tracing the effects of those changes in our corpus of practical certainties.[5]

---

[4]op. cit.

[5]This approach to deductive counterfactuals has been pursued, for example, by Peter Gardenfors, "Conditionals and Changes of Belief," in N. Niiniluoto and R. Tuomela (eds.), *The Logic and Epistemology of Scientific Change*, Reidel, Dordrecht, 1978, pp. 381-404; Alchourron, Gardenfors, and David Makinson, "On the Logic of Theory Change," *Journal of Symbolic Logic* **50**, 1985, 510-530.

Causal Connections

Probabilistic connections, at least of the sort mentioned first in the preceding paragraph, are easy to reflect counterfactually or hypothetically, since we may often simply add the required statements to our evidential corpus without worrying about deleting statements. In the case of serious uniform causal connections, the problem is complicated by the fact that almost always some deletions will have to be made, and that brings up the problems of vagueness and intention alluded to at the end of the last section. On the other hand, however, uniform causal connections, considered both counterfactually and hypothetically, are exactly the connections we need to take account of in playing our lives and building our tiger traps.

It is these uniform connections that best exemplify, if anything does, the mysterious powers and dispositions and affinities of scholastic philosophy. The dormative power of opium was made famous by Molière; the disposition of common salt to dissolve in water is a favorite illustration of twentieth-century philosophers; and the affinity of postage stamps for envelopes will serve to illustrate the third alternative.

We know that common salt is soluble. It must, of course, be put in water in order for it to display this disposition. There are constraints on the water, too: it must not be saturated with salt, the amount of water must be relatively large compared to the amount of salt, and so on. But assuming that our evidential corpus is appropriate and does not imply the denial of any of these constraints, if we add the sentence "S (the sample of common salt) is placed in B (the beaker of fresh water) and stirred" to our evidential corpus, then we will find the sentence "B dissolves" in our corpus of practical certainties.

It is clear that the process is not so simple as it is sometimes made out to be. The disposition of common salt to dissolve is rather constrained and hesitant; circumstances have to be just right. But it is nevertheless a disposition we can make use of, since we often know in advance that the conditions under which the disposition will display itself are satisfied. More accurately: we often have reason to accept in our corpus of evidential certainties statements whose truth corresponds to the satisfaction of these conditions. So what the disposition of salt to dissolve amounts to, at this level, is just the fact that we can often be practically certain that a particular bit of salt will dissolve.

Even more to the point: we can bring about the truth of those antecedent conditions; in fact, we can dissolve the salt if we want to.

At a more sophisticated level, there is more to it than this. We can understand the disposition of salt to dissolve in virtue of our understanding of the structure of sodium chloride and water. I suggested earlier that in large part this understanding is linguistic. I still maintain this, though what is required to account for the solubility of salt in water is not just the qualitative part of the atomic theory, but a quantitative account of the intramolecular forces in the salt crystal and their relation to the attractive forces of the atoms in water molecules. The quantitative part of the account is a matter of measurement (direct or indirect). Our knowledge of these quantitative constraints need not be terribly exact in order for us to feel satisfied that we understand what is going on. It is only at this higher level that we have a relationship between salt and water that might be referred to as the disposition of salt to dissolve rather than a mere uniformity; but at this level, it is not a qualitative disposition but a quantitative (and rather deep) fact.

Opium puts people to sleep in virtue of its dormative power. That sounds like a silly thing to say, though it is removed from the disposition of salt to dissolve only by the fact that we have a better understanding of the underlying quantitative mechanisms in the case of salt than in the case of opium. Nevertheless, it is useful. If we are concerned to preserve meat or pickles, knowledge about the solubility of salt can come in handy, whether or not we have any idea as to its mechanism. Similarly, knowledge about the dormative virtue of opium can come in handy if we need to put someone to sleep. Here is the connection to postage stamps and envelopes: if we want to mail an envelope, we put a postage stamp on it. Almost all mailed envelopes have postage stamps; but if there is a disposition, it is ours and social.

It has been suggested that both the solubility of salt and the dormativity of opium could be construed as promissory notes: to refer to these properties as "dispositional" is to claim that some day there will be available a detailed quantitative account of the mechanism through which the dispositions in question are expressed. Clearly, such notes do nothing to buttress the claim that we can make brine by adding salt to water. It is not clear that the promissory note provides anything but a pious and empty affirmation of the reign of universal causality: we may not know what it is, but there exists a quantitative causal explanation of the uniformity signaled by the dispositional statement.

Ultimately, however, some dispositions can be no more than brute facts: not all dispositions can be explained in terms of other dispositions. And many dispositions can be (and are) employed directly, whether or not we understand their underlying mechanisms.

Dispositions and Counterfactuals

There is one more variation on these themes that we may consider. I have based all probabilities on known relative frequencies, though in fact I have given no explicit semantic interpretation for the % operator. Of course, since I have said that the language is extensional, it can only be interpreted as something like frequency.

But there are circumstances under which we may want to base probabilities on theoretical measures, rather than on anything that can be construed as an actual relative frequency in the world. These cases are not as easy to find as one might think. The usual example is the newly minted coin that is to be tossed but once and then destroyed: is not its relative frequency of heads either 1 or 0? Of course. But to judge the probability of heads on that toss, we take as the reference class the set of coin tosses in general, and not the set of tosses of *that* coin about which all we know (presumably) is that it exhibits a relative frequency of heads of 1 or 0.

Suppose that we construct the one and only icosahedron ever to be constructed in this history of the universe. We construct it so that it is exquisitely well balanced. Given our knowledge of kinematics, we have every reason to believe that very tiny changes in momenta will lead to different sides coming up when it is tossed. (We may or may not suppose that the outcome of a toss is "determined" by the hypothetical momenta and velocities of the toss, combined with the actual resiliencies, and so on, of the icosahedron and the surface on which it is tossed.)

Under these circumstances, we want to say that the propensity for the icosahedron to land with a given side up is $1/20$, however seldom it is actually tossed. We cannot, by hypothesis, base this probability on actual relative frequencies of icosahedra. But we can still use actual frequencies: it is part of our general knowledge that regular n-sided polyhedra of the sort being considered land with each side up $1/n$th of the time. It is hard to take the possibility of such an icosahedron as requiring the existence of nonactual frequencies.

We could consider an experiment, such as spinning a dial, that has an outcome that we would want to associate with an irrational probability. This can be true only in a reference class containing a denumerable number of elements. Here we have left the actual world and entered an ideal one. No actual mechanism can run a denumerable number of times; no actual measurement can yield an irrational number.

We have already considered idealizations. In an idealization, we suppose that some actual phenomenon exhibits behavior that approximates that characteristic of our idealization. There are no ideal gases, but the ideal gas law represents a state of affairs that can be approached, and approached in ways that we understand.

The ideal coin lands heads half the time, but no actual coin — almost no actual coin, the exceptions being due to accident — throughout its career lands heads exactly half the time. Nor should we suppose that in the history of the world (the universe?), tossed coins in general will turn out to have landed heads exactly half the time. We represent the ideal gas law, in extensional form, by means of the generalization that all ideal gases satisfy it (in itself an empty claim) together with a measure of how closely actual entities approximate ideal gases (which gives, overall, serious and useful empirical content to the law).

Similarly, the ideal coin lands heads exactly half the time, in part because it is tossed a denumerable number of times, but also in part because it is flipped in a certain ideal way, it is perfectly balanced, and so on. So we have pretty good ways of approaching ideality, and of telling when we are closer to it and when we are farther away from it. There are no new problems introduced by the ideal coin, the ideal measurement sequence (Gauss showed that this would exhibit a normal distribution of error), or the ideal psi function in quantum mechanics, except for the problem of expressing the disposition in respect of which the entity is ideal.

Ordinary coins land heads with a certain finite frequency; no coin is tossed forever. And so, the tosses of no coin are actually independent: if the coin is tossed a total of N times, then for every time it lands tails, there is one less opportunity for it to land heads. But it is part of the very nature of the *ideal* coin that it is tossed forever, and so its tosses can be independent. There is no actual frequency to talk about here.

Similar remarks apply to the ideal normally distributed quantity (or any continuously distributed quantity, for that matter). Such distributions can be exhibited only in populations that are at least denumerable in size, and we can't talk about relative frequencies in such populations. Since the populations are imaginary anyway, they can have any sorts of properties we want them to have. In particular, let us give them idealized counterparts of relative frequencies — that is, measures.

So the measure of heads in the infinite sequence of tosses of the ideal coin is one-half, and similarly for the other examples. The ideal coin and coin-tossing apparatus also have all those properties (symmetry and the like) that we can measure in real coins. We can find coins and situations that approach the ideal, and we can have good evidence that they do.

When you and I flip a coin, does our probability derive from the closeness of our coin flipping to the ideal coin flipping? I think not. We (collectively) have direct inductive, statistical evidence concerning the frequency of heads among tosses of ordinary coins, and, more important, we have the indirect evidence provided by the folklore of coin tossing. It is true that our coin may approximate an ideal coin, but the actual frequencies in the world have more of a bearing on my probabilities than the imaginary measures in an ideal world, because even significant departures from ideality (no coin is symmetrical, else we couldn't decide which side was heads!) have little bearing on the relative frequency of heads.

On the other hand, if we wish to speak of the disposition of a coin, or of some counterfactual probability, then the ideal probability — the measure — seems to be what we want (e.g., if this unique, nearly ideal icosahedron had been tossed n times, the probability is nearly p that it would come down with the one-spot up between j and k times). The disposition of the icosahedron to land with the one-spot up is just the measure of the tosses of its ideal counterpart with this property.

Similar considerations suggest that this might also be a way to construe the continuous probability distributions of quantum mechanics. So let us take an *ideal stochastic object* to be a denumerable sequence of identically and independently distributed random quantities produced by an ideal chance setup associated with a corresponding *real* object. Examples: a sequence of tosses of an ideal coin and tossing apparatus, a sequence of measurements of a given quantity performed with ideal instruments. Of course, there aren't any ideal stochastic objects, any more than there are any ideal gases. Since we know this, we know that any ideal stochastic object characterized by a binomial distribution will fly over the moon, just like all the ideal gases.

It is (accidentally) true that every ideal gas satisfies the relation $P^2V^3=nRT$, but it is true necessarily—it is a theorem that, it follows from the conventions of our language that—every ideal gas satisfies the relation $PV=nRT$. Therefore, if something comes close to being an ideal gas, what it comes close to satisfying is the latter relation rather than the former. And if something *were* an ideal gas, it *would* satisfy the latter relation rather than the former.

The same with ideal stochastic objects. To the extent that the energies of the individual molecules in a sample of gas are distributed as the corresponding quantities in the ideal stochastic object embodying the normal distribution with mean m and variance s^2, the fraction with energies between e^1 and e^2 will be approximately that given by the corresponding integral of the normal density. (We say this, appropriately and informatively, even though we know that the number of molecules in the sample is finite, and that the energy is not normally distributed in any case, since there are no negative energies.)

Similarly, any real coin wears out and becomes asymmetrical after a finite number of tosses. Any real coin is tossed only a finite number of times, and so the tosses are not independent: the sequence of tosses will include exactly m tails among n tosses, and thus if the coin lands heads on its first toss, the frequency of heads in the remainder of the sequence is only $(m-1)/n$. But if C *were* an ideal coin, then the probability that it would land heads on each of the next 10^{10} tosses is exactly $1/2$ to the power 10^{10}. Said of any real coin, this would be ridiculous. (And as we shall see in the next chapter, this is not a useless thing to notice!)

Note, incidentally, that our standard interpretation of probability applies. We know, in **K**, that the "proportion" of sets of 10^{10} tosses of the ideal coin with the property in question has the exact value $(1/2)$ to the power 10^{10}; the next set of 10^{10} tosses is a random member of this set, relative to **K**, with respect to the property in question, since in the counterfactual case we know about this object exactly what we are said to know about it. And so, the probability is as advertised and is a real probability.

To achieve this simple and satisfying result, we have had to construe the operator % rather more broadly than we might have been inclined to do earlier: we have had to take it to reflect the more general notion of measure, rather than merely finite frequency. Furthermore, measure itself, in some uses, has an ideal or theoretical character. But this need involve no more philosophical strain than the ideality of the ideal gas.

14

Decision Theory

Expectation

Our efforts to acquire scientific knowledge yield, we believe, useful knowledge about the world. It is no doubt often gratifying simply to know more, but a greater immediate benefit of knowing more is that, knowing the future, one can make wiser decisions, and knowing how the world works, one can mold it closer to one's desires. If the river will flood, you can stay on the right side of it, or you can build a bridge.

There is a simple and unified decision theory that, if only it applied to everything, would in principle solve all the formal decision problems we face. It is called "Bayesian" decision theory because it requires, as does Bayesian statistical inference, a set of probabilities defined over the states of the world. These prior probabilities are updated in the light of evidence according to Bayes' theorem. In addition, we suppose that the states of the world can be assigned cardinal utilities.[1] This decision theory is based entirely on the principle of maximizing expected utility. Our first task, therefore, is to consider expected utility.

We can define *"expected utility"* in the framework we have already considered. "Prior probabilities defined over all the statements of the world" and "a cardinal utility defined for all the states of the world" sound rather grandiose. But if we distinguish relatively few states of the world, and if the values involved are simple enough, it is not so implausible to suppose that we have these functions. A simple artifi-

---

[1]That is, utilities that are unique except for the selection of a unit. Where there are such utilities, and, if there are, how to get at them, is controversial. Dealing with these controversies would require a serious excursion into value theory, which is beyond the scope of our interest here, though, of course, it is relevant to decision theory.

cial example will illustrate both Bayesian decision theory and the computation of expected utility.

Suppose that we can choose between receiving a prize of $1.00 if the first ball drawn from a certain urn is black and receiving a prize of $1.00 if it is white. If the urn contains 75% black balls and 25% white balls, the expected value of the first choice is $1.00 times the probability that the ball will be black, plus $0.00 times the probability that the ball will be white. To compute this expected value, we need the probabilities. What is the probability that the first ball drawn will be black? By now it should be clear that we need something more than the knowledge that 75% of the balls in the urn are black. We need the epistemological fact that "the next ball to be drawn" is a random member of the set of balls in the urn, with respect to being black. If that is so, we compute easily that the expected value of the first choice is $.75 and that of the second is $.25.

What does this mean? It does not mean that if we make the first choice, we will profit by $.75; on the contrary, that's impossible. We will either receive $1.00 or nothing. Nor does it mean that if we make the first choice, we will do better than if we make the second choice; the next ball to be drawn may turn out to be white. But it is common to say that "in the long run," making the first choice will turn out to be more profitable.

If we construe "in the long run" as finite, this defense is threatened with uselessness or circularity. Suppose that we are given the choice in question n times, where n is large. We assume that, relative to our corpus of knowledge, "the next n balls" is a random member of the set of all sequences—repetitions allowed, to represent replacement—of n balls from the urn with respect to having any particular proportion of black balls.

We may compute the expected value, just as before, of making the first choice each time or making the second choice each time. The expected value of making the first choice each time will exceed the expected value of making the second choice each time. But whether the first procedure will in fact yield more than the second procedure depends on how those first n draws turn out. To refer to the *expected* value of adopting the rule of making the first choice for each of the n trials is obviously circular: the reference to the long run was supposed to justify the rule of maximizing expected utility in the first place.

Suppose that we construe the "long run" as infinite. We are in the realm of the ideal, and one may well begin by asking what that has to do with our decision at a particular point in space and time. But in the ideal framework, if the conditions of randomness are met, we can assign a probability of $1-[1.0,1.0]$—to the claim that the first choice will be more profitable. But this does not mean that the first choice, even in an ideal infinite sequence, *will* be more profitable. It is perfectly possible, in the stochastically ideal model of this problem, that in an infinite sequence of draws (with replacement) from the urn, all will result in white balls. The probability of this outcome is 0, but the possibility is real.

Of course, we can say that the expected average value—the limit of ½ times the sum the outcomes on each of the first n draws times the probability of those outcomes—of making the first choice an infinite number of times will exceed the expected average value of making the second choice an infinite number of times. So what? Why, if only we had the principle of maximizing expected utility, this high expectation would warrant making the first choice all the time over the long run. But that was the very principle we were trying to justify. We have again encountered circularity.

Let us return to the finite run consisting of n trials. Suppose that we knew for sure that 75% of those n trials would result in a black ball. How does that influence our choice regarding the first draw? Circularity arises again: the choice of how to bet on the first draw, given that we know the frequency on the first n draws, is just the same as the choice of how to bet on the first draw, given that we know the frequency of black balls in the urn.

Assuming that the right conditions of epistemological randomness are met, we can compute that the probability is (say) $[.997,.999]$ that we will make between $.70n$ and $.80n$ dollars if we make the first choice each time, while if the second choice is made, the probability is the same that we will make between $.20n$ and $.30n$ dollars. On the view that allows detachment of probable conclusions, these statements can simply be included in the corpus of level .99. It is then not merely "probable" that we will do better by making the first choice; it is a matter of practical certainty. The difference of which we can be practically certain is the difference between $.30n$ and $.70n-\$.40$ per trial.

This does apply to the (large) set of n trials. It provides justification for adopting the intuitively preferable procedure as a rule for a sequence of n trials. It still does not provide any justification—as Peirce saw clearly—for choosing in the single case. For that, we must depend on the fact that the single case may be plausibly construed as one of a long sequence of choices.

Furthermore, there is a very serious limitation to this justification that in fact bears significantly on the use of the principle of maximizing expectation. It is that expectations can only be taken over plausible probabilities—that is, probabilities that are not too high or too low. Suppose that the urn contained 1000 balls, only 1 of which was black, but that choosing black and getting a black ball would yield $1000. Should we calculate the expected value of that choice to be $1.00?

If the level of our practical corpus is .998, we may argue that the probability of getting a black ball is, relative to the evidential corpus we have described, .001; and therefore, that it is practically certain that we will *not* get a black ball. Relative to our corpus of practical certainties, the probability, and therefore the expected value, of getting a black ball is 0. We will return to this question later—especially in connection with the problem of choosing levels of acceptance to be discussed in the next chapter. We will not assume that the answer just suggested—that high and low probabilities become categorical assertions—is completely persuasive. But the example does suggest that very high probabilities and very low probabilities may require special treatment.[2]

Maximizing Expectation

Let us continue the example of the last section, now making it plain where the Bayesianism comes in. Suppose that there are two urns, one containing 60% black balls and one containing 40% black balls. If we know which urn we are dealing with, the simple maximization procedure of the preceding paragraph suffices, given the principle that we should maximize our expectations. But we may not know which urn we're dealing with. If not, then which color we should choose to bet on should depend on the probability that we have a 60% urn rather than a 40% urn. This is already Bayesian, since what makes the difference for

[2]In a forthcoming paper, "Re-Modeling Risk Aversion: A Comparison of Bernoullian and Rank Dependent Value Approaches," in G. von Furstenberg (ed.), *Acting Under Uncertainty*, Kluwer, Dordrecht (forthcoming), Lola Lopes carefully discusses a number of psychological phenomena related to these questions. Her focus is more closely tied to the question of utility than to that of probability, and it is empirical rather than normative, but it throws much light on our present concerns.

us may be taken to be the "prior probability" of having a 60% urn as opposed to a 40% urn.

But we can go further. Once we have a prior probability for having one urn or the other, we can use evidence to update that probability. In the case at hand, the probability of betting on a sample of a certain character, subject to the usual constraints regarding randomness, is determined once it is determined which urn we have. Therefore we may apply Bayes' theorem, writing $Prob(H/E) = (Prob(H) \times Prob(E/H))/Prob(E)$ to get an updated probability of hypothesis H, that we have the 60% urn.

The updated probability of hypothesis H, of course, is what we use to calculate the expectation of each course of action open to us, if we have the evidence E. If $U(X,A)$ is the utility of the course of action A under the circumstances X, and if $Prob(E,X)$ is the probability of getting evidence E, supposing that the circumstances X obtain, then the course of action A to follow is that which maximizes the expected value

$$(U(X,A) \times Prob(E,X) \times Prob(X))/Prob(E)$$

summed over all X. $Prob(E)$ is just the sum over all X of $Prob(E/X)$.

Several problems have been raised concerning this style of analysis. The most serious one is the question of where the prior probabilities come from and what constraints they satisfy. This is a very general problem, and we shall come back to it shortly.

A different kind of problem concerns the evidential relation between the acts that I perform and the states of the world that obtain. Suppose that what act A I perform can serve as evidence concerning the state of the world — $Prob(X/A) \neq Prob(X)$. Then we may be led to worry about whether, by acting differently, I can influence that state of the world.

A number of examples illustrating this problem have been suggested. Some of the most interesting ones have been offered by Alan Gibbard and William Harper.[3] A useful approximation to an example described by Gibbard and Harper follows: Jones and several others are competitors for a promotion in their company. They are so evenly

---

[3] Alan Gibbard and William Harper, "Counterfactuals and Two Kinds of Expected Utility," in C. Hooker, J. Leach, and E. McClennen (eds.), *Foundations and Applications of Decision Theory*, Reidel, Dordrecht, 1978, pp. 125-162.

matched in qualifications that, Jones learns, the one who gets the promotion will be the one who scores highest on a psychological ruthlessness test that they have each taken on Friday, and that has been scored, but whose results will not be known until Monday.

Meanwhile, Smith, a subordinate of Jones, has failed to meet his sales quota. Jones knows that there are special circumstances surrounding this failure, that if Smith is treated leniently the company will benefit in the long run, and even that he can persuade his superiors in the company of these facts. Nevertheless, it is quite clear that for Jones to fire Smith would provide *evidence* of his ruthlessness, and therefore give him *reason* to believe that he had done well on the ruthlessness test on which his promotion depends, and therefore (assuming that the numbers are right) that the *expected value* to Jones of firing Smith could be greater than the expected value of not firing him — even though that act cannot *causally* influence the results of the ruthlessness test on which the promotion depends.

Gibbard and Harper use such examples to argue for a distinction between two kinds of expected utility, one of which measures the efficacy of an act in bringing about a desirable state and the other of which measures the welcomeness of the news that that act is to be (has been) performed. It would lead us astray to pursue this distinction here, but it is worth noting that careful consideration of randomness can save good old Smith, as well as considerations of counterfactual efficacy.

Suppose that R is the reference class of which Jones is a random member, with respect to getting the high score on the ruthless test, relative to *his* body of knowledge. Suppose that it is also the right reference class with respect to our body of knowledge. Now Jones is confronted with the problem of whether or not to fire Smith. From our point of view, knowing that Jones did (or did not) fire Smith might lead us to a new reference class R' (e.g. rising young executives who celebrate a potential promotion by firing old friends) for establishing the probability that Jones scored well on the ruthlessness test. This is surely appropriate. It represents, to us, a new fact about Jones.

But does *Jones* learn something new about Jones by making the decision? Perhaps not — or not in general. Insofar as Jones is *choosing* between acts, he must assume that his choice and his score on the ruthlessness test are independent. That is to say, *for him, R* must still be the right reference class, else he would not be regarding his act as a result of his own free choice. He is still, for himself, a random member of R, and remains one despite the fact that he is considering the possibility of firing Smith. There *is* something strange about this, for Jones would be just like us in evaluating the probable ruthlessness of Brown, who is in a situation just like that of Jones and who does fire his Smith.

For Jones to *believe* that he has a choice is exactly for him to believe that there is no connection between the way he chooses and the score he made on the test. It is therefore for him to put himself in one of two subclasses of R, each characterized by the same distribution of ruthlessness. The two subclasses are (a) the intersection of R and an appropriate set corresponding to choosing to fire Smith and (b) the intersection of R and an appropriate set corresponding to choosing not to fire Smith. This is one of the ways in which the problem was structured in my "Acts and Conditional Probabilities."[4]

Here is yet another way to think of this situation. We are watching the world unfold. When we learn that Jones has fired Smith, we change the reference class for his ruthlessness from R to R'. We would quite properly pay him more to change jobs with us, since we would have more evidence that he had got the lucrative job for which he was a candidate. Or we would rationally change the odds at which we would bet that he got his promotion. *After* he has made his choice, after he has fired Smith, he is in the same position we are. He can reflect and think, "Well, maybe I did better on that exam than I thought." But he cannot do this in the process of deliberation. He can choose his act, but he cannot choose his reference class at the same time.

What is special about Jones' situation when he is calculating this expected value of the alternatives he faces with respect to Smith? Simply the fact that neither alternative is actual, and therefore that R remains the right reference class for him even as he is contemplating alternatives that, when actualized, would provide evidence that would change the appropriateness of R.

It thus seems that in circumstances of complete knowledge—that is, when all the relevant probabilities are based on reference classes about which we have precise statistical information—the Bayesian principle of maximizing expected utility serves quite generally. But this is not always, or even often, our situation.

[4]*Theory and Decision* 12, 1980, 149-171.

Interval Constraints

In general, of course, the probabilities that we feed into our computation of expected utility will not be point valued. In the simplest cases we may consider a finite number of alternatives, each of which has an interval-valued probability relative to our practical corpus. In more complicated cases, we may have a continuously varying quantity on which our utility depends, and we may know that quantity has one of a family F of distributions. In either case, we emerge with intervals of expected utilities corresponding to the contemplated acts.

The problem we must face is that there may be no act that maximizes expected utility. It might be that the utility interval associated with one act lay entirely above the utility interval associated with any other act — in which case we could say that it maximized expected utility. But this obviously need not be the case.

For example, consider the following urn problem. Let us suppose that we are offered the chance to draw a ball either from urn 1 or from urn 2. If we get a white ball, we receive $1.00, and if we get a black ball, we receive $5.00. To compute the expectations of the two acts — draw from urn 1 or draw from urn 2 — we need the probabilities of the two possible outcomes. Let us suppose that in either case the ball chosen will be a random member of the set of balls in the urn from which it is chosen. The probability of a white ball, in either case, will thus be the smallest interval in which we know the proportion of white balls to lie.

Suppose that we know that the urns contain only white and black balls, and that the first urn contains between 10% and 30% white balls, while the second urn contains between 30% and 70% white balls. The expected utility of choosing from the first urn is $.3 \times 1.00 + .7 \times 5.00$ at the least and $.1 \times 1.00 + .9 \times 5.00$ at the most — that is, in the interval [3.8, 4.6]. Similarly, from the second urn, the expected utility lies in [2.2, 3.0]. No matter which among the possible distributions actually obtains, the expectation of drawing from the first urn exceeds that of drawing from the second urn.

But the situation would be different if what we knew about the contents of the urns were different. Suppose that what we know is that between 40% and 60% of the balls in the first urn are white, and that our knowledge about the second urn remains the same. Then the expected utility of choosing the first urn is [2.6, 3.4] and that of choosing the second urn is still [2.2, 3.0]. In this situation it is possible that we will do better (in the long run) by choosing the second urn, and it is possible that we will do better (in the long run) by choosing the first urn. Talk of the "long run" is only a way of talking, of course: we get to choose once, and we merely characterize our choice in terms of the long-run properties of making such choices in general.

We can also consider situations in which the interval corresponding to one choice is properly included in the interval corresponding to the other. All told, we may consider three situations:

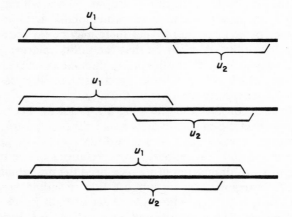

These situations are reminiscent of the possibilities we considered in basing our choice of a reference class on our knowledge of relative frequencies. In fact, the problem is altogether different. We have settled on our reference classes, and therefore on our probabilities, and what we need is to know how to weigh utility intervals.

Choices of the sort illustrated in the first picture are easy. If under all possibilities consonant with our knowledge act A is preferable in expected utility to act B, then surely act A is preferable. This principle represents a plausible first step in a lexicographic decision procedure. We can generalize it slightly: we can say that one act dominates another if its expected utility is at least as great as the expected utility of the other regardless of what possible state of nature, among those consistent with our knowledge, may actually obtain, and greater in at least one state. Thus the option of choosing to receive a prize if a red or a white ball is drawn from a rainbow urn containing many colors of balls dominates the option of choosing to receive a prize if a white ball is drawn. Whatever the proportions of white and red balls, if each color is represented, there have to be more balls that are either white or red than there are that are just white. But what is the next step, if any?

One possibility is a maximin strategy: choose so as to maximize the minimal expected utility. In instances like the second or third illustrated above, this strategy would recommend choosing that act whose minimum expected utility was the greatest: u_2 in the second example and u_2 in the third example.

A different possibility is to adopt some way of combining the upper and lower expected utilities of an act to obtain a single expectation. This amounts to treating utility as a linear combination — presumably the same one for each act — of the upper and lower expected utilities. The simplest case: take the average of the upper and lower utilities.

It is also possible to be very optimistic and choose that option with the highest upper expected utility — a maximax option.

In general, where our interval probabilities come from confidence methods, it will be the case that the higher the level of the rational corpus containing a statistical statement about a given subject matter, the broader the interval or the broader the bounds of the distributions we may accept as characterizing that subject matter. For example, if the evidence we have is such as to warrant accepting in the corpus of level .9 that between .5 and .7 of the balls are white, it might give us warrant to accept only an interval of [.45,.75] in the corpus of level .95, but it might give us warrant to accept an interval of [.55,.65] in the corpus of level .8. It is possible that by considering a lower-level corpus, we can get a decision between two courses of action by the dominance principle alone.[5]

Finally, it is possible that we should take the constraints of rationality to have been exhausted with the dominance principle. Any option not ruled out by the dominance principle, we might feel, can have something to say for itself under circumstances we have no reason to suppose impossible. Some grounds for thinking that this might be right are provided by cases in which there are a lot of options "in the middle" and an extreme option that would be chosen by the minimax strategy and an extreme option that would be chosen by the maximax strategy. Cases like this can be imagined where rejecting the extremes would seem reasonable; others, formally similar, in which minimax

---

[5]This approach is being actively pursued by Ronald P. Loui; see "Interval Based Decisions for Reasoning Systems," in J. Lemmer and J. Kanal (eds.) *Uncertainty in Artificial Intelligence*, North Holland, Amsterdam, 1986, pp. 459–472.

seems reasonable; and yet others in which maximax seems reasonable. On the other hand, this may be a reflection of the fact that the description of the situation uses numerical terminology, which may be intuitively translated (illegitimately) into probabilities.

In any event, the dominance principle does seem relatively clear. It has been questioned,[6] but to examine the arguments that have been raised against it would lead us too far astray. And other constraints on rational choice are not so persuasive.

Utility

Although we have noted that value theory is beyond the scope of our concerns, there are down-to-earth questions of value to which science may be applicable. For example, if the decision problem is how to build a bridge that meets certain specifications, and no aesthetic or social considerations obtrude, then the utility of a design is the negative of its cost in coin of the realm. Cost effectiveness can be measured.

Matters are not generally quite this simple, though, since the "specifications" themselves are open to modification. We may demand an expected life of 50 years and find that for a tiny increment in cost, we can get an expected life of 75 years. Or we may find that 45 years is easy but that 50 years is twice as expensive. Furthermore, we may not be concerned only with the expected life, but with other features of the distribution of longevity in the class of objects of which it is a random member, such as the variance. If two items have the same expected life, but the variance characterizing the distribution to which one belongs is much larger than the variance of the distribution to which the other belongs, the latter may offer significant advantages.

In general, though, we must balance variance against expectation, and there is no standard way of doing this. If you have a choice between two light bulbs (at the same cost), each with the same variance of life, but one of them has a longer expected life, it is clear that the one with the longer expected life (fashion aside!) is preferable. It is almost as clear that if you have a choice between two light bulbs, each with the same expected life, but one of them has a much smaller

---

[6]Edward McClennen, "Sure-Thing Doubts," in B. P. Stigum and F. Wenstop (eds.), *Foundations of Utility and Risk Theory with Applications*, Reidel, Dordrecht, 1983, pp. 117–136.

variance of life expectancy, then it is the one with the smaller variance that is usually preferable for reasons of predictability. (We know when the latter will burn out; i.e., we can be *practically certain* that we will have to replace it between d_1 and d_2, while in the case of the former, our *practical certainty* covers a wide range of dates. Alternatively, in minimax terms, we can be *practically certain* that the latter will last for more than h_2 hours, and we can be *practically certain* that the former—with the smaller variance—will last for more than h_1 hours, but $h_1 > h_2$.)

Utility theory, at least in its present form, and particularly in an interval version of its present form, gives us no way of balancing these desiderata against each other. It may be that there are compelling arguments for doing the balancing in certain way, but they have not yet been discovered.

On the other hand, quite different issues are raised by questions of aesthetics, questions that concern the relative value of present and future utilities, questions that concern "incommensurable" values. Which is preferable, a beautiful bridge that lasts for 40 years or a less beautiful bridge that lasts for 75 years? Or a beautiful bridge for k dollars as opposed to a slightly less beautiful bridge for somewhat less than k dollars?

In one sense, these questions raise deep philosophical issues; in another sense, they do not. We can look at the choices people make (in a free market) and infer a dollar difference between a light bulb (or a bridge) with an expected life of n hours (years) as opposed to an expected life of m hours (years). There is no real difficulty here, any more than there is in ascertaining how much extra people are willing to pay for fresh eggs. It is a straightforward question of measuring people's preferences.

The same approach—that of analyzing people's preferences among alternatives—can also give rise to measures in dollars of aesthetic worth (of bridges) and even of the values—admittedly very hard to evaluate directly—of health and life. Assuming that people are relatively rational (ideal agents), we can find out not only what they value, but what its exchange rate is in dollars.

Surely there are difficulties with such a program? Of course. Nobody ever said it was easy. But in principle it is a measurement problem, and we know in general how to do measurement problems. Naturally, as in other measurements, we obtain only intervals as acceptable characterizations; but this creates no new problems, because we are already committed to intervals through our conception of probability.

Is this all there is to the matter? Not at all! We may know what Smith values, but intelligibly say that his values, though consistent, are unreasonable. (Perhaps he is some kind of religious fanatic.) In the same way, it may be argued that a group of people, though their collective values are coherent, are collectively unreasonable in their evaluations. In fact, it suffices that it is open to question the utility assignment of an individual or a group of people in order for us to raise the quite different question of what values an individual or a group of people *ought* to hold.

Furthermore, this sort of question is not idle. We know that it is possible to change the utility function of an individual (e.g., by philosophical argument) or of a group (e.g., by an advertising campaign). Each of us has preferences about the preferences of other individuals and groups, and each of us has some power to influence those preferences by example, argument, or persuasion. The question of how effective a given form of influence is is an engineering question. The deeper question, of what your preferences and values and commitments *ought* to be, is, at base, a philosophical question.

Prima facie, then, there is a place of fundamental importance for value theory. At least, if it is intelligible to question the measured values of any individual or group, it is an open possibility that there is an objective standard against which any empirically discovered set of values can be weighed. It is at this point that we get essentially beyond the scope of the philosophy of science, if indeed our intuitions about the intelligibility of questioning empirically discovered values are correct.

Nevertheless, it is worth noting that in much of our deliberation concerning values, it is not any deep question of value theory that is at issue, but a straightforward computation of expected utility in terms that, if not agreed upon, at least depend only on the measurement of reasonable preferences and the computation of expectations. And further, it is worth noting that we can often come to conclusions on the basis of very rough and limited assignments of utility. (But, of course, we can't even *talk* to people who assign infinite positive or negative utilities to anything!) We need not solve all the philosophical problems of value theory in order to design a bridge within a given set of social and practical and even aesthetic constraints.

Puzzles

There are some well-known puzzles concerning decision theory. First, there is a group of puzzles that depend on the question of whether or not one should take one's own choice to be evidence. These are the puzzles that we have already dealt with, of which Gibbard and Harper have provided the paradigmatic examples. These puzzles are dealt with by taking seriously the reality of free will from the point of view of the agent.

Second, there are puzzles, due to Allais and Ellsberg, that appear to show that people systematically violate the principle of maximizing expected utility.

Here is an example of the kind of puzzle considered by Ellsberg[7]: You are given an urn containing 30 red balls and 60 balls that are white or blue; an *unknown* proportion w of the total are white. You are to choose between options A and B, with the following payoffs:

	Red	White	Blue
A	100	0	0
B	0	100	0

By a straightforward calculation, since w is unknown, the expected value of A is the interval $[100/3, 100/3]$ and the expected value of B is the interval $[0, 200/3]$. Expected utility doesn't dictate which alternative you should take, but if you add a principle to the effect that you should maximize your minimum gain, choice A seems to be preferable.

Now consider, in the same situation, the following pair of choices:

	Red	White	Blue
C	100	0	100
D	0	100	100

Your expected utility calculation now yields $[100/3, 300/3]$ for choice C and $[200/3, 200/3]$ for choice D. Again, expected utility doesn't solve

---

[7]Daniel Ellsberg, "Risk, Ambiguity, and the Savage Axioms," *Quarterly Journal of Economics* **75**, 1961, 528–557.

your problem, but maximizing your minimum gain would now lead you to choose D.

Not only do these choices seem intuitively reasonable, but in fact, many people do choose in just this way. So what's paradoxical? The puzzle is that according to Bayesian principles, one should make choice D if and only if one makes choice B. For any given value of w, the (point-valued) expectation of B exceeds that of A if and only if the expectation of D exceeds that of C. Thus, if we represent your uncertainty by a point-valued function, whatever its value may be, there is no way in which you can rationally both prefer A to B and D to C:

$$100/3 > 100\ w \text{ if and only if } 100/3 + (1-w)100 > 200/6$$

The upshot of the puzzle, it seems to me, is that point-valued uncertainties don't provide the right input for all decision problems. In particular, when our knowledge of proportions is interval-valued, we should take our expectations to be interval-valued, and the Bayesian principle of maximizing expected utility doesn't always give a determinate decision. Here we have supplemented the principle of dominance (leave out of consideration actions whose maximum expected utility is less than the minimum expected utility of some alternative) by a *minimax* principle: among undominated options, choose one whose maximum loss (minimum gain) is a minimum (maximum). Perhaps this principle is not always the next one to be used; the point is that the principle of maximizing expected utility is not sufficient for all decisions.

The paradox of Allais[8] runs as follows. The chance setup in this case is completely known: you face an urn containing 100 balls: 89 white, 10 blue, and 1 red. Here is the first pair of options:

	Red	White	Blue
E	1,000,000	1,000,000	1,000,000
F	0	1,000,000	5,000,000

Many people find that E is the natural choice.

[8] M. Allais, "The Foundations of a Positive Theory of Choice Involving Risk and a Criticism of the Postulates and Axioms of the American School," in M. Allais and O. Hagen (eds.), *Expected Utility Hypotheses and the Allais Paradox*, Reidel, Dordrecht, 1979, pp. 27–148.

Here is the second pair of options:

	Red	White	Blue
G	1,000,000	0	1,000,000
H	0	0	5,000,000

Here many people find that H is the natural choice.

What is paradoxical here is that however you value (or disvalue!) a million dollars or five million dollars, the choices of E over F, and of H over G are inconsistent. Let your utility for a million be u_1 and your utility for five million be u_2. Under the Bayesian analysis, your choice of E over F says that:

$$u_1 > 0.89\, u_1 + 0.01\, u_2$$

from which it follows that $u_1 > u_2$. But your choice of H over G says that

$$.01\, u_2 > .11\, u_1$$

from which it follows that $u_1 < u_2$.

Here the situation is not that resulting from what is called "indeterminacy of probability"[9] (though even in the preceding case, there was nothing indeterminate about the *probability* — that was determined exactly to be [1/3, 2/3]). One way to deal with this problem is in terms of the marginal utility of money.[10] More complicated analyses are possible,[11] but there is also the possibility of saying, from a normative point of view, that people are *taken in* by this example, just as people, in general, are wrong in estimating how many people must be considered before you have a better than even chance that two of those being considered have the same birthday. This is strongly suggested by the fact that if the middle column in the two-choice matrices is deleted (on the grounds that in *that* case it doesn't matter which alternative you choose), you have exactly the same alternatives in the two cases, and so should choose the same way.

---

[9] Isaac Levi, "On Indeterminate Probabilities," *Journal of Philosophy* 71, 1974, 391–418.

[10] This is the approach of Isaac Levi in "The Paradoxes of Allais and Ellsberg," *Economics and Philosophy* 2, 1986, 23–53.

[11] For a complete psychological analysis with experimental data, see Lola L. Lopes, op. cit.

In any event, there seems to be one clear principle of rationality: that if the value of outcomes is measured in utilities, one should not choose an outcome whose maximum expected utility is less than the minimum expected utility of some other outcome. The probabilities entering into this alternative should not be greater than the level of the practical corpus relative to which we are making the calculation. It is silly to take a probability of 10^{-10} seriously if you are taking .90 as your level of practical certainty; relative to that corpus, such events just don't occur.

This principle does not provide for decision among all actions, though it may eliminate a lot. In order to choose among the alternatives that remain, some further principles will be required. One such principle is the minimax principle mentioned above. Whether or not this is to be regarded as a *principle of rationality* seems open to question. It does seem pretty compelling. But it isn't clear that it would compel universal agreement. In addition, there are related questions concerning utility. It seems pretty clear that the relationship between money and utility is not linear, but how far one can or should go in cooking up complicated nonlinear relations between money and utility is unclear. Finally, it is also unclear where to draw the line between rational and irrational behavior. We don't want a normative theory of rationality according to which most people are irrational most of the time; on the other hand, a normative theory that merely reflects how people actually choose would be vacuous. We seek guidance, not from popularity polls, but from thought and analysis.

15

Levels of Corpora

Full Belief

One problem that may have been nagging the reader for a long time is the problem of choosing levels of rational corpora. Since what goes into a corpus is what has a probability higher than the index of that corpus (relative to a corpus of higher index), that index has a bearing on what (and how much) is in a corpus. Furthermore, we have *two* levels to deal with, since we are concerned both with the evidential corpus and with the practical corpus. What principles can we employ to select these levels?

A useful way to approach this question is through the analysis of full belief, or acceptance. What is it to *accept* a statement? Partial belief is characterized (at least by Bayesians) in terms of a propensity to make or fail to make bets. Given a range of dollars that satisfies the elementary axioms of utility, we can say that I have a degree of belief equal to one-half in the statement that the next toss of this coin will land heads in virtue of my propensity to bet. Of course, amounts of dollars only approximately satisfy the axioms of utility; and, of course, I may be indifferent between accepting and rejecting a bet over a nonzero range of odds. But these are just the classical difficulties of measurement.

Can we give an analysis of full belief along the lines of the proposed analysis of partial belief? Is to have full belief in S to be willing to pay a unit of utility for a ticket that will pay a unit of utility if S is true? Or to be indifferent between a ticket that pays a prize P in any case and one that pays a prize P only if S is true? If it is logically possible that S should be false, one would think the natural response would be: why take the chance?

There is a fair amount of empirical data to the effect that when dealing with events whose relative frequencies are close to 0 or 1, people's choices between alternative gambles no longer seem to fit the general pattern of maximizing expected subjective utility.[1] These deviations suggest that in the experimental situation, events with frequencies that are very close to 1 are simply assumed to occur, and events with relative frequencies very close to 0 are simply assumed not to occur. This leads to conflict with the Bayesian model of deliberation, but that is not my concern here.

What I am concerned with is the fact that if "full belief" is to be considered a limiting case of "partial belief," then we should suppose that partial beliefs progress in an orderly way toward this limit. Empirically, this does not seem to be the case. To be sure, our object is a theory of *rational* belief, rather than an empirical descriptive theory of belief, and we *might* say that people tend to be irrational about events with extremely high or extremely low relative frequencies. But we should not conclude with unseemly haste that other people uniformly leap into irrationality under the same uniform circumstances. It may be that there is an intuitively acceptable notion of rationality within which this general tendency can be accounted for, and perhaps even regarded as rational.

R. B. Braithwaite[2] offers an alternative and, at first sight, more useful dispositional interpretation of full belief. It is (roughly) that a person has full belief in S if he acts as if S were true. The difficulty is that whether or not a person "acts as if S were true" depends on what is at stake. A person may act as if a certain vaccine is nontoxic when it comes to vaccinating monkeys, but not when it comes to vaccinating children.[3] Braithwaite requires that full belief in S be represented as a

---

[1] See Ward Edwards, "Subjective Probabilities Inferred from Decisions," *Psychological Review* **69**, 1972, pp. 109–135. and references therein; see also C. Stael von Holstein (ed.), *The Concept of Probability in Psychological Experiments*, Dordrecht, Reidel, 1974, and D. Kahneman, P. Slovic, and A. Tversky (eds.), *Judgement Under Uncertainty. Heuristics and Biases*, Cambridge University Press, Cambridge, 1982. For a recent discussion, see Lola L. Lopes, "Re-Modeling Risk Aversion: A Comparison of Bernoullian and Rank Dependent Value Approaches," forthcoming in G. von Furstenberg (ed.), *Acting Under Uncertainty*, Kluwer, Dordrecht.

[2] R. B. Braithwaite, "Belief and Action," *Proceedings of the Aristotelian Society,* **XX**, 1946, 1–19.

[3] The example comes from Isaac Levi and Sidney Morgenbesser, "Belief and Disposition," *American Philosophical Quarterly* **1**, 1964, 1–12.

disposition to act as if *S* were true under any circumstances to which the truth of *S* is relevant. But as Levi and Morgenbesser have pointed out, "for any contingent proposition *p* on which action can be taken, there is at least one objective relative to which a nonsuicidal, rational agent would refuse to act as if *p* were true. Consider, for example, the following gamble on the truth of *p*: if the agent bets on *p* and *p* is true, he wins some paltry prize, and if *p* is false he forfeits his life. However, if he bets on not-*p*, he stands to win or lose some minor stake. . . . Hence . . . the agent could not rationally and sincerely believe that *p* where *p* is contingent."[4]

Levi and Morgenbesser go on to consider more complicated reconstructions, in which actions are taken to depend on circumstances, motives, and stimulus, as well as beliefs, and show that what is involved is more like a promissory book than a promissory note. Our concern here, however, is not with the general problem of construing beliefs dispositionally—a very knotty problem indeed—but with the far simpler and more limited problem of making sense of full belief *presupposing* an understanding of such parameters as motives, stimuli, external circumstances, and the like. In particular, we shall presuppose what is itself transparently (I should say, rather, "opaquely") dispositional: a cardinal utility function for the agent over states of the world.

Let us begin by looking more closely at the matter of "acting as if" *S* were true. If I bet at even money on heads on the fall of a coin, it might appear as if I were acting as if "The coin will fall heads" were true. The appearance is misleading. My *action* is that of making a bet, and making a bet is not acting as if the proposition that is the subject of the bet were true. On the contrary, it is acting as if a certain relative frequency or propensity characterized the *kind* of event at issue. Thus betting at even money on heads *is* acting as if at least half of the tosses of coins yielded heads. This much, at least, seems relatively straightforward, though in detail we would have to take account of my aversion to risk or my pleasure in excitement. In fact, we can no doubt characterize my betting behavior concerning coin tosses in general by saying that in *ordinary circumstances* I act as if the relative frequency or propensity of heads were one-half, or, more precisely, as if the distribution of heads among coin tosses were binomial with $p = 1/2$.

---

[4]Ibid., 3.

But this claim concerning the relative frequency or propensity of heads is surely a contingent claim. Would you risk your life in exchange for a paltry prize on the proposition that the propensity of this coin to yield heads was one-half? Or even "close" to one-half? Or even on the proposition that the propensity of coins in general to land heads was one-half? Hardly.

But we can easily enough imagine bets and stakes concerning the long-run behavior of the coin that would strike us as reasonable. (I'll give you odds of 10 to 1 that if we flip the coin until it wears out, the relative frequency of heads will end up between .48 and .52.) Now, of course, I am "acting as if" almost all coins and flippings are symmetrical enough to yield that result.

On the other hand, let us suppose that I am willing to act as if "half of the tosses yield heads" is true—that is, to bet at even money on heads on a toss of a coin. From "half of the tosses of coins yield heads," it follows that (let us suppose) 1/1,346,451 of the sets of 10,000 tosses fail to yield a relative frequency of heads between 4000 and 6000.[5] (The exact numbers are irrelevant.) Then should I not be willing to offer or receive odds of $1.00 to $1,346,452.00 against this outcome in a particular case? It seems to me that I should not.

One might think that this is a reflection of the size of the stake involved—1 million seems large. This feeling is reinforced by the example of the possibly toxic vaccine: so much more is at stake in inoculating children than in inoculating monkeys that we should have much more evidence that the vaccine is nontoxic before we inoculate children than before we inoculate monkeys.

But I think that this intuition is mistaken, and that we are being misled by our background knowledge of the frequencies of disease and the function of vaccination. Suppose that it is not a new vaccine at issue, but a new antibiotic, which is demonstrably effective against disease D in pigs. Humans are also subject to D (though only rarely); but whereas pigs rarely fail to recover from this disease, humans almost always succumb to it. We have some evidence regarding the toxicity of this drug in both humans and pigs; to fix our ideas, suppose that we can say that the probability that it is nontoxic to pigs and the probability that it is nontoxic to humans is about the same—say, about .80. It seems quite clear that one would give the drug to a group of

---

[5] I refer here literally to a *set* of tosses, not a *sequence* of 10,000 tosses; hence there is no requirement of "independence."

children known to have D, but that one would not give it to a group of pigs known to have D. That is, with respect to the children, one would act as if the drug were nontoxic, despite the high stakes involved, but with respect to the pigs, one would not act as if the drug were nontoxic.

This example suggests that what is crucial in whether or not a person "acts as if S were true" is not the total magnitude of the stake involved but the ratio of the amount risked to the amount gained. Let us take this as our basic idea: A fully believes S in the sense of the ratio (r/s) just in case, in any situation in which the ratio of risk to reward is less than $r:s$, A simply acts as if S were true. Symmetrically, A will then also fully disbelieve $\sim S$—that is, in any situation in which the ratio of the risk of counting on $\sim S$ to the benefit of counting on S is greater than $s:r$, A will simply act as if $\sim S$ were false.

An example will help to make this clear. Suppose that I fully (99/1) believe that my car will start this afternoon, but that I don't fully (r/s) believe it for $r/s > 99/1$. Suppose that act a is an act that will cost me \$50.00 if my car doesn't start, but otherwise will yield a benefit worth \$1.00. Since $50/1 < 99/1$, I will perform act a—that is, act as if I knew that the car would start.

In general, if the stakes are in the range $1:99$ to $99:1$, I will simply act as if I knew for sure that the car would start.

Full r/s *Belief*

Let us try to formalize full *r/s* belief. The idea behind our formalization is that if S is a factual sentence, there may be some circumstances under which A will believe the sentence S, but there may also be circumstances under which A will *not* believe S, and not believe the negation of S. If not too much is at stake, A will believe (accept) S; otherwise, A will not accept S. The following principle captures this idea:

> B1 A fully (r/s) believes S if and only if for every action that A takes to cost r' if S is false and to yield s' if S is true, where $r'/s' < r/s$, A acts as if S were true.

A number of consequences of B1 follow:

> T1   For any action that A takes to cost s' if $\sim S$ is false, and to yield r' if $\sim S$ is true, where $s'/r' > s/r$, A acts as if $\sim S$ were false.

This is the other half of the intuitive characterization of acceptance in the previous section. To see that it follows from B1, note that to cost s' is to yield $-s'$, and to yield r' is to cost $-r'$. For $\sim S$ to be false is exactly for S to be true. B1 then says that if $-r'/-s' < r/s$, A should act as if S were true. But $-r'/-s' < r/s$ if and only if $s'/r' > s/r$.

    T2 A fully (r/s) believes S if and only if there is a $p\in[0,1]$ such that A fully $p/(1-p)$ believes S.

This is so because, if $p=r/(r+s)$, then $r'/s' < r/s$ if and only if $r'/s' < p/(1-p)$; p is the probability of S in classical Bayesian terms, and we are contemplating no odds that would justify a bet against S.

    T3   If A fully (r/s) believes S, and $1 < r'/s' < r/s$, then A fully (r'/s') believes S.

A cannot act both as if S were true and as if $\sim S$ were true, though, of course, A can simultaneously *bet* on both S and $\sim S$; indeed, this is just how bookmakers make their living. From this it follows that if $r/s < 1$, A cannot fully (r/s) believe S.

    T4   If A cannot both act as if S is true and act as if $\sim S$ is true, then if A fully (r/s) believes S, $r/s > 1$.

    T5   If A is logically omniscient and $S \vdash S'$, then if A fully (r/s) believes S, A fully (r/s) believes S'.

Clearly, it need not be the case that if A fully (r/s) believes S and fully (r/s) believes S', then A fully (r/s) believes $S\&S'$. The ratio of the stakes involved in acting as if both S and S' were true may no longer be less than r/s.[6]

Suppose that P is a classical epistemic probability function defined relative to what A fully (r/s) believes. We are assuming here that probability is a logical relation of the sort we have considered elsewhere. By calling it a "logical relation," we mean that the probability of S relative to any set of other statements is logically determinate (it won't be bothersome if there are a few peculiar circumstances under which it is not defined), and has a certain value regardless of what anyone whose body of full beliefs corresponds to that set of evidence statements does believe or would believe.

---

[6]Since $r'/s' < r/s$ and $r''/s'' < r/s$ imply $(r'+r'')/(s'+s'') < (r/s)$, one might think not. But the risk is in fact $r' + r'' + r'''$, where r''' is the additional cost of both S and S' being false.

B2 If $Prob(S)=[p,q]$, and $q<r/(r+s)=p'$ and Bd is the act of paying d for a unit of return contingent on S, then A will rationally perform Bd if $d<p$; A will rationally refrain from Bd if $q<d$; and if $p\leq d\leq q$, then the rationality of A cannot be faulted whether or not he performs Bd.

So far as it goes, this principle conforms to the principle of maximizing expected utility. But now suppose that $p\geq p'=r/(r+s)$. Then according to the generally accepted scheme of maximizing expected utility, the previous analysis should still apply: if $p>d$, it is worth buying a ticket for d that will return one unit if S is true, even if $d>p'$. But this is equivalent to staking a possible gain of $1-d$ against a possible loss of d on the truth of S. This is not in the range of stakes contemplated in the sense of "full (r/s) belief" unless $d/(1-d)<r/s=p'/(1-p')$. Since $d>p'$, $d/(1-d)<r/s$ is impossible, and the ratio of stakes in Bd is not among those contemplated. Similar considerations govern a bet against S.

Thus the conventional Bayesian wisdom breaks down. If we are interpreting full belief as full (r/s) belief, then statements having probabilities greater than $p'=r/(r+s)$ or smaller than $1-p'=s/(r+s)$ are *not* fit subjects for bets.

But statements whose probabilities relative to the set of A's full (r/s) beliefs are greater than p' — there may well be some — are not necessarily fit subjects for full (r/s) belief, either.

This appears to represent a gap in our normative theory of full and partial belief. A statement S may qualify neither for full (r/s) belief nor for full (r/s) disbelief, nor be such that the Bayesian principle of maximizing expected utility always gives us guidance for actions depending on S. But we can fill this gap, and at the same time achieve a unification and generalization of our theory, by recalling that the parameter r/s is adjustable. Thus S, which presents a problem relative to full (r/s) belief, may not present a problem relative to a higher or lower level of full belief.

Of course, in view of the fact that evidential probabilities are interval-valued, we already have a gap in our decision theory. We supposed above that P was a "classical epistemic probability" — that is, real-valued. We can make the connection between real-valued P and evidential probability *Prob* in the following way: we say that a classical real-valued P *conforms* to the probability function *Prob* (with regard to a set of sentences) just in case for every sentence S in the set, $P(S)\in$

$Prob(S)$. An action whose outcome depends on S is rationally permissible if it is rational according to some conforming P function; it is rationally obligatory if it is rational according to every conforming P function; it is irrational if it is rational according to no conforming P function. So there are already bets that it is rational to take and also rational to refuse, even before we consider these special cases in which the probabilities are so high as to come close to our acceptance level.

Let us write the index r/s in the normalized form $p'/(1-p')$ or, better yet, replace it by the single number $p' = r/(r+s)$. The number p' corresponds naturally to the *level* of our corpus of practical certainties. We have said that a statement should get into the corpus of practical certainties by being probable enough—having a lower probability at least equal to p'—relative to the evidential corpus. The following argument suggests (but doesn't quite *demand*) that the relation between the indices of the evidential and practical corpora be that the practical corpus be indexed by the square of the index of the evidential corpus. For, suppose that S_1 is in the evidential corpus. Then for any sentence S, $S \rightleftarrows S \& S_1$ will be in the evidential corpus in view of the theoremhood of

$$S_1 \rightarrow (S \rightleftarrows S \& S_1)$$

Thus, if S is in the practical corpus, so will $S \& S_1$ be in the practical corpus. What this means in general is that if S_1 and S_2 are in the evidential corpus, their conjunction will be in the practical corpus. Since this fact depends on no relation between S_1 and S_2, it will hold for S_1 and S_2 that are independent stochastic events whose probabilities are exactly p. Their conjunction will have the probability p^2 relative to the same corpus that gives them each a probability of p, and so, since their conjunction is practically certain, that suggests that p' should equal p^2.

Call a sentence S anomalous for full p' belief if it is neither a fit subject for bets in accordance with B2 nor a fit subject for full p' belief in accordance with B1. If S is anomalous for full p' belief, then it should be acceptable as worthy of full q' belief, where $q' < p'$. And at some higher level, $r' > p'$, it may be a fit subject for Bayesian guidance. All that is required is to make full p' belief depend on having a probability of at least p' relative to the corpus of knowledge whose index is $p'^{1/2}$. More precisely, we adopt a third principle of rationality (we now write "full (p)" for "full $(p/(1-p))$"):

B3 *A* should give full (*p*) belief to *S* if and only if the minimum probability of *S*, relative to *A*'s full ($p^{1/2}$) beliefs is greater than *p*, or *S* is obtained by observation, and relative to *A*'s full (*p*) metabeliefs, the maximum probability that *S* is in error is less than $1-p$.[7]

This principle leads immediately to the result that by shifting the ratio *r/s* slightly, we can always resolve the anomalies in at least one way:

T6 If *S* is anomalous relative to *A*'s full (*p*) beliefs — that is, if its minimum probability is greater than *p*, but it is not a member of *A*'s full (*p*) beliefs — then *A* should give full (*p'*) belief to *S*, where $½ < p' < p$. (This holds in virtue of theorem T3, which requires that $p > ½$.)

It is also possible to resolve anomalies the other way — that is, to move to a level *p'* such that the Bayesian maxim does apply. This need not be the case when we consider interval probabilities, but it can be shown to hold for conforming real-valued epistemic probability functions, and these are admissible as a basis for bets.

T7 If *S* is anomalous relative to *A*'s full (*p*) beliefs, then there exists a $p' > p$ such that relative to *A*'s full (*p'*) beliefs, the probability of *S* is less than *p'*, and therefore the Bayesian maxim is appropriate for bets on *S* relative to *A*'s full (*p'*) beliefs.

Suppose that there is no such *p'*. Then, relative to every one of *A*'s sets of full (*p'*) beliefs, where $p' > p$, *S* has a probability greater than or equal to *p'*. In particular, this applies to *A*'s set of full (*p*) beliefs — that is, *A*'s beliefs concerning mathematical and logical truths. But then there is a *p'* — namely, 1 — such that the probability of *S* relative to *A*'s full (*p'*) beliefs is greater than *p*, where *p'* is greater than *p*. But this contradicts the assumption that *S* not be fully (*p*) believed by *A*.

That this doesn't work for interval-valued probabilities can be seen as follows. Suppose that relative to *A*'s full (*p*) beliefs, the probability of *S* is $[p',q']$, where $p' < p < q'$. It need not be the case that there is a $p^* > p$ such that the upper probability of *S* is less than p^* relative

---

[7]This clause suggests sensibly that it is possible to come to fully (*p*) believe something on the basis of observation without having to stake one's life or one's honor on it; it is embodied in our earlier treatment of observation and measurement.

to A's full (p^*) beliefs. Thus there are conforming P-functions according to which we can bet on S, but not all conforming P-functions need have this property.

Constraints on Full and Partial Belief

Let us see how this framework can be taken to impose constraints on rational belief and rational degrees of belief. One of the relevant factors in the dispositional analysis of belief, however we go from there, is the set of circumstances in which the agent finds himself. But although this is a relevant factor, it is not one that should lead us to despair. We do not want to say that we can provide a dispositional analysis of rational belief only if we know the circumstances of the agent in infinite detail. Indeed, this would preclude our being able to assess the rationality of others or to improve the rationality of ourselves, for we can never articulate our circumstances in infinite detail.

What is required is that we be able to characterize a broad class of circumstances under which our analysis is to apply. Within the framework suggested here, this class of circumstances is characterized precisely by the range of ratios of stakes that the agent has (implicitly) in mind. It is easy enough to alter the circumstances so that they lie outside this range—at least hypothetically. This is precisely what Levi and Morgenbesser are doing when they point out that, whatever the evidence for S, there exist circumstances under which the agent A will not act as if S were true—for example, when his honor is on the line against a paltry prize. But this is precisely because this represents a ratio of stakes outside the range implicitly and initially contemplated by A. Often this shift can be accomplished relatively easily and realistically by means of the simple query: "Wanna bet?"

Within the framework we have been considering, we could consider a notion of full belief in which the ratio of stakes was very close to unity—full ($1/2+\epsilon$) belief. This is just what is done sometimes by epistemologists who require of S merely that it be "more probable than not" in order to be worthy of belief. But this is not a very interesting sense of "full belief." Ordinarily we want our beliefs to remain fixed over a relatively wide range of circumstances—that is, to be suitable for a relatively wide range of ratios of stakes. A range from 10 : 1 to 1 : 10 might seem more plausible.

Now the actual stakes, in any circumstance, depend on the agent's utility function: therefore so also does their range. The set of circumstances relative to which our analysis is to be performed should thus be represented by a function of both the agent A's utility function (a function of A himself) and a ratio of stakes: $C(A,r/s)$.

Consider full belief first. We say that A has full (r/s) belief in S, just in case for any circumstance $c \in C(A,r/s)$, A acts as if S were true. Note that for A to act as if S were true is not for A to perform any particular action (in any ordinary sense of "particular"), as Levi and Morgenbesser seem to suggest.[8] This is not the place to attempt a characterization, but it nevertheless seems reasonable just to say that there is a certain class of "deliberate behaviors" that are ruled out by "A's acting as if S were true" (e.g., betting against S if the opportunity is offered) and a certain class that are required (e.g., betting on S if the opportunity is offered). The existence of these classes of deliberate behaviors seems to be all that is needed to give content to the notion of "acting as if S were true."

A's full (r/s) belief in S is rational, according to the framework principle B3, just in case the probability of S, relative to A's full (p) set of beliefs, is at least p' where $p^2 = p'$, and A's full (p) beliefs are themselves rational in turn. This raises an obvious problem to which we shall return shortly.

Now let us consider A's partial (r/s) belief in S. If A's degree of belief in S is to be characterized by a real number q, it will be exactly that number q such that if A is *compelled* to take one side or the other of a bet at odds of $q : 1-q$ on S, he will be indifferent as to which side he takes. But this seems unnatural and foreign to the notion of the set of ordinary circumstances $C(A,r/s)$; we should seek a gentler characterization of degree of partial belief. We can get at this by supposing that there is a range of ratios, say from $q : 1-q$ to $q : 1-q'$ such that A would be indifferent about taking either side of the bet. (Remember that the bets are in utilities, so A's enthusiasm for gambling and his reluctance to take chances are presumably already taken into account.) A's partial degree of belief then comes to be characterized by the interval $[q,q']$. Put another way, if he is offered the opportunity to bet against S at odds less than $(1-q'):q'$, he will take it; and if he is offered the opportunity to bet in favor of S at odds less than $q:(1-q)$, he will take it.

---

[8]Levi and Morgenbesser, op. cit., 6, fn. 30.

Now when is A's partial (r/s) belief in S rational? Clearly, when, relative to A's body of full (p') rational beliefs, the probability of S is no narrower than $[q,q']$. This means that under any circumstances $c \in C(A,r/s)$, A ought to bet on S at odds better than $q:(1-q)$ and ought to bet against S at odds lower than $(1-q'):q'$.

Note, however, that if the odds do not lie in the range $[r:s,s:r]$, then A is *not* in the circumstances $C(A,r/s)$ envisaged by the analysis. Suppose that A is offered a bet on S at odds between $q:(1-q)$ and $q':(1-q')$. Then he is under no rational constraint either to accept or to reject the bet. But this is all right. We still have perfectly good characterizations of A's partial (r/s) belief and of the constraints that this belief must satisfy in order to be rational.

There are two additional points to be noted. If, relative to A's rational full (r/s) beliefs, the probability of S is $[p,q]$, where p is greater than $r/(r+s)$, then for no $c \in C(A,r/s)$ could A rationally bet against S. But if S is not a member of A's full (r/s) beliefs, which it may not be, neither can we demand that A act as if S were true under any $c \in C(A,r/s)$. What should A's doxastic attitude toward S be? It seems perfectly natural to say both that S is not a statement that A believes, and also that it is not a statement against which he would bet under any circumstances in $C(A,r/s)$. This seems to be a perfectly natural (and indeed familiar) situation to be in. But we may also suppose that the possibility of a *serious* bet on S would change the circumstances contemplated from $C(A,r/s)$ to $C(A,r'/s')$, where $r'/s' > r/s$, and the odds of this possible serious bet are such that they fall in the range $r':s'$ to $s':r'$. Then the previous analysis would hold.

Another oddity arises if, relative to A's full (r/s) beliefs, the probability of S is $[p,q]$, where $p < r/(r+s) < q$. Then A should bet on S at odds less than $p:(1-p)$, but there are no odds at which he should bet against S. (Note that he should not be ruled *irrational* should he bet against S at odds corresponding to probabilities in the interval from p to $r/(r+s)$.)

Further Consequences

This approach has a number of consequences for real life that bear on relatively practical matters. We all have useful stochastic knowledge; trivially, that well-maintained gambling apparatus is very nearly fair. From such knowledge we may infer that stochastically ideal apparatus produces certain outcomes with very small (or very large) probabilities. Thus the probability of heads on each of 1000 tosses of a fair (ideal) coin is $1/2^{1000}$. There is nothing wrong with this computation. But on the basis of the preceding analysis, it is only useful for computing expectations in a set of circumstances that includes an enormous range of odds: $1-1/2^{1000} : 1/2^{1000}$ to $1/2^{1000} : 1-1/2^{1000}$. Such a range of odds does not characterize anybody's practical concerns, even leaving aside the fact that no coin could be known to be ideal in the required strong sense.

This explains, at least in part, why nobody wants to play the St. Petersburg game for what it is "worth." (The St. Petersburg game is one in which player A offers to pay player B a prize of 2^k dollars if a head first appears on toss k in a sequence of coin tosses. It is easy to compute that the value of this game to B is infinite. Given any range of feasible odds, we can calculate the value of the corresponding truncated St. Petersburg game; but for any reasonable set of full beliefs, almost all of the possible outcomes of the St. Petersburg game should be fully disbelieved.

Similar considerations bear on our proper epistemic attitudes toward very rare events. In my practical corpus of level $1-\epsilon$, for example, it is simply not credible than an event of probability less than ϵ can occur. Therefore the expected value of a prize contingent on such an event is exactly 0, regardless of how valuable the prize is in itself. Correspondingly, in a corpus of level $1-\epsilon$, the expected cost of a disaster that has a probability of less than ϵ is 0, however dreadful the disaster.

Of course, to contemplate a glorious prize (eternal joy) or a horrible disaster (eternal damnation) may lead the agent to increase the range of risks and benefits—the range of odds—he wishes to take account of in his corpus of practical certainties. But observe that this has the following effect: it constrains the agent to operate with a very high standard of evidence (the square root of the level of the corpus of practical certainties that corresponds to this range of odds), and that means that the actual content of his corpus will be sparse. The ideal coin lands heads 1000 times in a row exactly $1/2^{1000}$ of the time; but no real coin is ideal, and if our standard of practical certainty is given by the index $1-1/2^{1000}$, we cannot claim to know very much about the real world. More generally, in order to use a probability of ϵ in the computation of an expectation, the probability of the statistical law on which the first probability is based must itself be greater than $1-\epsilon$, and only relative to an evidential corpus of a level higher than $1-\epsilon$ does that computation make sense in this system. A possibility, relative to a corpus of level $1-\epsilon$, whose probability is less than ϵ is not a real possibility.

Given knowledge of an agent's preferential dispositions, we can characterize the sets of circumstances $C(A,r/s)$. Then, given a knowledge of the agent's full (r/s) rational beliefs, we can divide the actions he might contemplate into those he ought rationally to perform, those he ought rationally to refrain from, and those concerning which there are no rational constraints either way.

If we wish to assess the rationality of his full (r/s) beliefs, we may do so by considering the probabilities of the statements fully (r/s) believed, relative to the agent's full (r'/s') beliefs, where $(r'/s')^{1/2} = (r/s)$. We may do this for any set of circumstances $C(A,r/s)$.

This serves to give us a complete handle on both the agent's rational beliefs and his rational actions, subject only to the characterization of his utility function—obviously a nontrivial matter, and one that itself involves dispositional interpretation.

Thus, although the dispositional characterization of actual belief may seem to open up a Pandora's box of confederate notes, and thus to represent an obscure research program rather than an enlightenment, the dispositional characterization of rational choice calls for only one blank check (to be filled out in A's utilities) and otherwise admits of relatively clear-cut prescriptions, provided that the epistemic characterization of rational belief previously outlined is accepted.

As in many other areas of endeavor (in geometry, for example), it turns out to be easier to prescribe the ideal (the rational) than to describe the real (the actual). But the ideal is not without relevance to the real, as we have already noted in connection with ideal descriptive theories. The constraints on rationality are those we should strive to embody.

16

Speculation

Speculation and Creativity

Speculation is often contrasted with scientific knowledge. The existence of life on other planets of the solar system is "sheer speculation," while the chemical nature of the atmospheres of the other planets is a matter of scientific knowledge. As the very example suggests, the line is not sharp. In the present chapter we will not dispute the distinction, at least as a relative one, but we will examine the importance to science of relatively unbridled speculation.

We have been supposing all along that scientific knowledge itself is rational. I should emphasize that what I mean here is scientific *knowledge*; leaving truth to one side, let us take this to be justified empirical belief. We have argued that justification in this sense is mechanical. By this is meant two things: (a) statistical inference is mechanical and (b) the choice between languages is mechanical.

To call statistical inference mechanical is to claim that, given a language of the regimented sort we have been discussing, given a body of evidence, given a level of acceptance, and given background knowledge that is itself justified, an algorithm can yield the results of inference.

To say that the choice between languages is mechanical is to claim that the procedure outlined earlier for comparing the novel empirical content of two bodies of practical certainties can be mechanized and will yield a determinate outcome.

Suppose that we are speaking a given language. Given a set of input data, there is exactly one state of rational belief that is justified by those data. More explicitly, if the data enter the corpus of level p, then the set of sentences acceptable at level p^2 is exactly determined.

Similarly, given two languages L_1 and L_2, given a certain body of experience, we can apply the criteria suggested for choosing between the languages. These criteria, too, are intended to be purely mechanical.

We have considered the suggestion that they may consist of adding certain observation reports to the Ur-metacorpus (of each language) and then counting the predictive observational statements that appear in the practical corpora corresponding to the two languages. Of course, this is very simple-minded and naive. In languages that include measurement functions — as any interesting language does — we require a more sophisticated technique of measuring novel predictive information. And we have to settle on what statements to add and in what numbers. But these matters do not seem to be matters of great principle; it is hard to imagine serious disagreements about the details.

Where one may well expect serious disagreement is about the whole idea. It reduces scientific judgment to a matter of counting or measuring in a language. Choosing between languages, as we have considered it, like performing statistical inferences from given data, sounds pretty pedestrian. Where is the romance and wonder of science?

Some of the romance — the nonpedestrian, creative part of science — lies in speculation. Speculation is often thought to be not quite respectable compared to serious science. It doesn't answer to canons of rationality and responsibility. Anybody is free to speculate about little green men on Mars. Speculation is seen as being a lot like story-telling.

But this is not to say that speculation does not play a vital role in scientific thought. It is one thing to have rational, computational grounds for assigning probabilities, or for choosing between given languages, or for making statistical inferences in a given language from given statistical data and with given background knowledge. It is quite another to *have* the languages to choose between, to decide what data to *get*. It is in this regard that speculation plays an important role.

None of this can be claimed without argument. Some writers (e.g., N. R. Hanson[1]) believe that there is a genuine logic of scientific discovery, though many still follow the old positivist line and distinguish between the context of discovery and the context of justification, reserving only the latter as grist for the logic mill. But even in the context of discovery there seem to be distinctions to be made. Surely, one thinks, there is a difference between speculating about little green men

---

[1] N. R. Hanson, *Patterns of Discovery*, Cambridge University Press, Cambridge, 1958.

and speculating about a spiral ladder structure for the DNA molecule. It may not be easy to characterize the difference, but surely there is one.

There is another extreme view according to which all of science is relatively unconstrained, except for principles that themselves are subject to shifts in fashion and popularity. This view is associated (with less than perfect justification) with T. S. Kuhn and P. Feyerabend.[2] It is sometimes thought that this approach reduces the philosophy of science to the history or sociology of science—one might say, the sociology of speculation.

In the sections that follow, we shall examine a bit more carefully what should be understood by scientific speculation; what our attitude toward it should be; and what sort of rational constraints it is subject to.

The Role of Speculation

The most important role of speculation is that of providing us with scientific theories. Recall that it has been claimed that the generalizations, laws, and axioms that characterize a scientific theory are best construed as linguistic constraints on the language of science: they characterize our scientific language, and the role of observation and experiment is merely to offer a probabilistic assessment of the soundness, reliability, accuracy, and so on of judgments made using that language. To be warranted in adopting a particular language is to have reason to prefer that language to some alternative scientific language. In some cases (classical thermodynamics, for example), the axioms and laws are explicitly about *ideal* entities and processes. As such, they clearly cannot be falsified—there are no objects to falsify them—and construing them as a priori linguistic constraints of the language is clearly appropriate.

Where do such constraints come from? How do we get to invent new languages embodying new constraints? By speculating; by telling stories. It is true that we have statistical grounds for believing that

---

[2]The best-known sources are T. S. Kuhn, *The Structure of Scientific Revolutions*, University of Chicago Press, Chicago, 1962, and P. Feyerabend, "Problems of Empiricism," in R. G. Colodny (ed.), *Beyond the Edge of Certainty*, University of Pittsburgh Press, Pittsburgh, 1965, pp. 145-260. Kuhn's work has been especially influential in the past 15 or 20 years.

almost all ravens are black, let us say; why not simply suppose that all—literally all—ravens are black and see what it does to our corpus of practical certainties? Or, assuming that we have some reasonable measure of ideality, we might take it to be a feature of our language that all *ideal* ravens are black. As we observed in an earlier chapter, such a language, in this simple case, might offer us no advantages over the more phenomenalistic language in which we can accept only the statistical generalization that almost all ravens are black. But in a richer context that need not always be the case. For example, we might have an evolutionary theory according to which all species of birds that fall in the general class to which ravens belong are such that all *members* of the species are naturally black. "Naturally black" allows for sports and pots of paint.

Let us consider a more complicated story. For example, suppose that there is an imaginary, ideal relation of being longer than, and an imaginary, ideal operation of concatenation, that satisfy the standard measurement axioms. Clearly, the observable relation that we can judge directly, and the concatenation operation that we perform on physical objects with our own two hands, do not at all satisfy these axioms. So the ideal story is a wonderful fairy tale. But in this case, there are enormous benefits to be gleaned in the corpus of practical certainties.

Far more important than that, though, are the benefits accruing to a *collection* of people who all believe in the foot rule: you can measure your refrigerator, report the result of the measurement to me, and I can tell (with practical certainty only), by measuring my door, whether or not your refrigerator is wider than my door—a matter that may be of considerable practical importance. Less frivolously: we can standardize our bolts and screws and engine parts. All industry depends on the possibility of indirect and secondhand comparisons of length.

Consider Kepler's planetary laws. Kepler was convinced that there was an important relation between the orbits of the planets and regular solids. Why not? If you're just telling a story, why not make it a good one? But, of course, there is a difference between *telling* a story and *believing* that the story you've told is true. And there is a difference between according the story *rational* belief and according it *mere* belief. When we speak of scientific knowledge, we are speaking of the story we have reason to believe. And we never did, nor did Kepler, have reason to believe the story that involved the regular solids. But he did, and we do, have reason to believe the story about each planet sweeping out equal areas in equal times (ideally!).

This distinction corresponds to the distinction made early in the days of logical empiricism, already referred to, between the context of discovery and the context of justification. The distinction was then applied to what were considered substantive empirical hypotheses rather than linguistic conventions. The idea was that anybody could, without any sort of justification, *propose* a law or generalization: there was considered to be no "logic" of discovery, and if inductive logic were a logic of discovery, no inductive logic. But given a generalization, it was felt that the question of whether or not it was justified was a question that was open to objective adjudication. Perhaps by reference to logical probability, *à la* Carnap; perhaps in only one direction, falsification, *à la* Popper.

Our situation is a bit different, since we do not regard laws and theories and generalizations as having (in themselves) empirical content. Nevertheless, we have offered a sketch of what is claimed to be an objective standard for choosing between languages; in that regard, the context of justification admits of an objective adjudication.

Corresponding to "discovery," however, we have speculation. While there were no constraints on the laws and generalizations that might be proposed on the old view, it often seemed that the *language* in which such laws and generalizations might be expressed was to be taken as a *given*. One of the big problems for these views was the problem of theoretical terms: how could they be introduced into language, how could they be related to ordinary observational terms (or defined away), how could they be expressed in a language lacking them?

On the view being proposed here, the whole language is fair game for speculation. We can add terms freely—not only theoretical terms, but observational terms as well. We can impose whatever a priori constraints on those terms may strike our fancy. Generalizations, laws, axioms of any sort can be adopted as part of the language. The freedom is complete. But, of course, the freedom to imagine stories is useless to someone with no imagination. And that is why, even on my very pedestrian view of justification, imagination and creativity are essential to science.

We have been talking as if the process of testing were completely global. It is. But the fact that it is a global process—that we have to choose between *whole* languages—does not mean that the languages have to be *completely* different. In fact, a small change—one that does not have large consequences in the predictive observational content of our practical corpus—can perhaps be evaluated quite easily. A small change in *axioms*, however, may yield a rather large change in predictive observational content. So it may not always be an easy matter to compare two languages (to decide if a suggested change is advantageous).

Polysemous Tolerance

There is no need to be intolerant of any speculation. There is all the room in the world for any number of good stories. But most scientific speculation is not offered merely for entertainment; it often—but not always—comes with some claim of plausibility, reasonableness, or truth. In line with the standards for choice between languages we considered earlier, of any two languages one must be regarded as superior to the other, or they must be regarded as equally good. I doubt that anything near the second case arises in real life.[3]

If ties are unlikely or impossible, then only one language of any finite number can be rationally adopted. There exists scientific *knowledge*, embodied in the constraints characterizing the best language (of the ones we have available), and there is also *mere* scientific (or "unscientific") speculation embodied in the constraints of other languages. But doesn't this rather put a damper on pursuing scientific change? If it is not reasonable to believe in a new theory, can it be reasonable to look for the evidence that will make it reasonable to believe it?

In one sense, it is not. Before the evidence is in, it is not rational to believe in the new theory (speak the new language). Remember that the evidence bears on the statistical distribution of error associated with the new theory. There *may* be circumstances under which it is not rational to spend the time and effort to develop a theory, evaluate its essential constants, and so on, unless it is at least possible that it should become rational to believe the theory.

But people are not always altogether rational. And the observation we have just made provides one more reason (among many others!) to

---

[3]Possible claim to the contrary: weren't wave and corpuscle theories of light both useful at the same time? Sure, but as we observed earlier, there is no difficulty in adopting an instrumental language in which both aspects of light are captured: under certain circumstances, light acts as if . . .

be glad of this. It may not be quite rational, in the sense of communicable or arguable, to spend time, money, and effort looking for evidence for a theory that it is not (yet) rational to believe. But it can be a fascinating gamble.

Consider a speculator, A. She tells herself, and her colleagues, if any, and perhaps the funding agencies, a story. We assume that it is not the accepted story, and that it is not immediately warranted by the evidence that the community already has. A may believe her story — that is, she *may* think (irrationally) that the language corresponding to her story actually *is* preferable to the currently superior language. Or she may quite rationally believe (hope) that the language corresponding to the theory *may* in due course be shown to be preferable to the currently accepted language. A's irrational belief may quite rationally inspire her to devote time and effort to looking for the evidence that will show this. This may be no more than mild eccentricity.

There is no reason that A's colleagues, as well as the rest of society, should not tolerate this mild eccentricity. But matters change when A wants to use social resources, and not merely her own free time, to pursue the evidence for this theory. To justify the use of social resources, we want some (statistical) evidence to the effect that A's eccentricities have paid off before, or that eccentricities like A's have paid off. We may ask: of what class of eccentricities is A's eccentricity a random member, and what range of success frequencies is characteristic of that class? And then we (society) are in a position to ask: what benefits if she is right? What costs if she is wrong? It may still be worth investing in A's eccentricity, even if the payoff rate is rather low. It is hard to see anything evil about *this* application of cost-benefit analysis to scientific research.

Investigating the little green men on Mars is not likely to pan out; investigating a new conjectured structure for a certain molecule is the sort of thing society could well support, even if 90% of such investigations don't yield definitive results.

To come down to earth, to more realistic issues, we can easily afford to be tolerant of the irrational biological beliefs of those who accept the biblical creation story as opposed to the evolutionary story, just as we can afford to be tolerant of a person who offers his own story to account for a certain event of extinction within the classical evolutionary framework. We have a lot more reason to support the research of the latter than of the former. And we have an absolutely clear-cut and categorical reason to keep the former away from our children: this person wants to present a speculation as a rational belief. This is a form of prevarication that cannot be tolerated in a civilized society.

In sum: scientific orthodoxy does exist. Where it exists, it determines, on rational, logical grounds, the language and content of our practical corpus of rational beliefs. But it is neither complete nor the final word. Speculations serve, potentially, to fill out details or to replace the current orthodoxy.

From a private point of view, we can afford to be very tolerant of speculation; and we can afford to be tolerant even of those who irrationally believe their own speculations. When we are being asked to make a personal or social investment in another person's speculation, it is reasonable for us to evaluate that investment in the same sorts of terms we use to evaluate any investment: past performance, potential yield and so on—just the questions that referees for the National Science Foundation are asked.

Finally, to be asked to tolerate false claims as if they were *justified* when they are not—claims that theory/language X is superior to theory/language Y when that is not the case—is categorically beyond the pale.

Standards

But are there no objective, logical standards by which we can compare the reasonableness of the speculation that there are little green men on Mars and the speculation that there is yet another subatomic particle? Maybe not. But that does not mean that there are no standards at all. Perhaps they are not standards of reasonableness or rationality, but standards of taste. I have referred to scientific (and unscientific and antiscientific) speculation as "stories." There was design behind the diction.

There are evaluative standards that we do bring to bear on stories. They are literary and aesthetic standards, and there is less than universal agreement about what they are, but they are standards nonetheless. Unity, coherence, and simplicity are among the desiderata; intrinsic plausibility might be. Ad hocness, artificiality, and the like are defects. Not only do such standards apply to stories, scientific or otherwise, but they apply to poetry, music, painting, and all the other arts. But how could such aesthetic standards be construed as "logical"?

One talks, or hears talk, of the logic of poetry, of music, and even of moving picture shows. Does this mean *logic*? *Real* logic? Of course not. It is a metaphor meant to convey that each of these disciplines has an inner structure that has been worked out over long periods of time within a traditional but changing framework. The tradition reflects the analogy captured by the metaphor; the fact that it has changed as the discipline has developed reflects the fact that it is *only* a metaphor.

Viewed this way, we may speak of the logic of scientific speculation, where "logic" is used in this same metaphorical sense. There is an inner structure to scientific speculation and a tradition. But it is a tradition that gradually changes over time.

Traditional metaphysics is a rich source of scientific speculation. Democritus and his atoms, for example, are not seriously different, as speculation, from Dalton and his. What sets Dalton apart is the fact that he could *do* something with his atoms. They could be connected, through a structure of probabilistic and uncertain knowledge, a statistical matrix of distributions of error, to predictive observational certainties. It is not, of course, the same speculation in the two cases. Democritus' atoms are not at all Dalton's atoms. The point is that, considered independently of the evidence that allows Dalton's speculation to have consequences in the practical corpus, both are abstract scientific speculations of a strikingly similar character.

A natural question to ask is whether these aesthetic qualities that provide standards for scientific speculation—not exactly the same standards we would apply to poetry or music—are evidence for the truth of the theories/languages they rate as superior, or for their acceptability, or are portents of their ultimate acceptability.

There certainly appears to be no connection between satisfying certain aesthetic standards and being true. In fact, we have rarely used the term "true" in this volume, and it is not clear that it serves a useful function in the evaluation of scientific theories. We know—or can claim to know—when a sentence is rationally acceptable, that is, when it appears in a certain rational corpus based on a certain set of observation *reports*. We can know this even of observation *sentences*. But on the reconstruction of scientific knowledge we have been considering, there is no way to legitimately claim to know that a sentence is true (except that to accept it may be glossed as *claiming* that it is true), even when the sentence is an observation sentence. Truth has entered into our discussion in only a very modest way.

Is there a link between aesthetic quality and acceptability? Again, it seems that the answer is negative, except for the fact that people tend only to tell and appreciate pretty stories, and what we accept must have been told as a story at some point. Kepler's notion that the orbits of the planets were linked to the regular polyhedra surely has wonderful aesthetic qualities to recommend it, but that hardly made it true. Nor, I think, can a case be made, even roughly, that aesthetic quality is a portent of future acceptability. Aesthetic quality may be a necessary condition in the sense that an ugly theory/language will never receive serious consideration.

Thus, if there is a connection, I suggest that it runs from acceptance to beauty. Any theory that is acceptable, at any point in time, is going to have its share of the aesthetic virtues appropriate to scientific speculation, as those virtues are understood at that time. The reason is not far to seek. Who wants to create an ugly language/theory? Even if the most effective scientific language were ugly, nobody would invent it, and thus we would never know about it.

This emphasis on aesthetic qualities may be misleading in one regard. Many people assume that there are no objective standards of aesthetic quality, so it may be assumed that we are claiming that scientific taste is just as subjective as literary taste. It is easy to overdo the emphasis on taste. To be sure, in some sense preference for one scientific theory over another, as *speculation*, may be a matter of taste. But here as elsewhere, it is possible to distinguish *good* taste from *bad* taste. And when it comes to preferring one scientific theory to another as a matter of evidential support — that, we have being trying to argue, can be made a highly objective matter.

17
The Limits of Science

Marginal Matters

It has been amply demonstrated over the past few hundred years that science provides immensely powerful tools for altering the natural world. Most people would agree that it has also provided understanding and explanation. But it is not at all clear that the tools and methodology can lead to resolution of all of our cognitive quandaries.

It has been alleged, for example, that scientific knowledge and religious knowledge are so different in character that they cannot even come into conflict. But the recent debates about evolution and creation suggest that perhaps they can come into conflict after all. It has been argued that the rift between matters of fact and matters of value is so deep that scientific knowledge of matters of fact can have no bearing on our knowledge, if knowledge it be, of matters of value. But it has also been claimed that "ought" implies "can"; and if science yields "can't," that should therefore, by *tollendo tollens*, cancel the obligation of the "ought."

Closer to home, there are two issues underlying science itself that are alleged not to admit of any sort of scientific resolution. One is the issue of realism and instrumentalism: is there any conceivable way in which the tools of science can be brought to bear on this ontological question (if ontological question it be)? The other is the question of the presuppositions or postulates of the scientific method. It has been argued that we require substantive empirical assumptions, postulates, or presuppositions — ultimate presuppositions — before we can claim soundness for our scientific arguments. If this is so, these presuppositions themselves cannot be defended by scientific argument. The very foundations of science itself may lie beyond the limits of scientific inquiry.

Even if some of these claims concerning the limitations of science are true, they are, as just stated, both cryptic and vague. In what follows I shall expand on them a bit, and I shall argue that in most respects they are defeatist, or self-serving, or arbitrary, or some combination thereof. One way of doing this, of course, would be to argue that all cognitive activity was, *per definitionem*, "scientific." It would follow, all too glibly, that insofar as the questions at issue were cognitive, they would admit of "scientific" treatment. Therefore let us begin by offering a rough and ready characterization of what we should take to be "scientific," and of the "natural world" to which we apply science. Rather than finding this characterization so loose as to be question begging, you will probably find in it a shockingly narrow, straitlaced, and old-fashioned empiricism. Despite the obituaries that have appeared in some of our leading journals, it isn't clear that empiricism is dead.

"Empiricism," roughly, is the doctrine that what we know, we know on the basis of experience. "Experience" is a pretty loose term, too, but in general, we can take experience to arise out of interaction between the experiencer and the natural world. By the natural world, I mean, first of all, the objects and events of ordinary experience: sticks and stones, trees and flowers, birds and beasts; the succession of day and night, of the seasons; growth and decay; the ordinary behavior of animals and people. Ordinary experience is somewhat vague; the ordinary experience of the Eskimo is not that of the South Sea islander; and the ordinary experience of the astrophysicist is different yet. But they still have much in common.

I take this ordinary experience to be crystallized in ordinary language and its dialects, and in common judgments, and these common judgments to lie at the foundations of science and indeed of all knowledge. I also take them to be the judgments that science tries to allow us to anticipate. The predictive content of our corpus of practical certainties is written largely in the language of commonplace observational judgment.

There are two ingredients in the development of scientific knowledge. Both have their roots in common sense and ordinary responses to the world. One ingredient is "induction," by which I mean statistical inference. If the frequency of A's among a large number of B's has been observed to be f, infer (other things being equal), with an appropriate degree of confidence, that the frequency of A's among B's in general is approximately f. More generally: if the quantity Q is found to be distributed according to the function F in a large sample of B's then infer (other things being equal), with an appropriate degree of confidence, that Q is distributed in B's in general approximately according to F.

This form of inference has yet to receive its full articulation, perhaps, though I would claim that the basic ideas can be found in earlier works.[1] In particular it is not easy to specify what "other things being equal" means or how to determine an "appropriate degree of confidence" — but serious efforts to lay bare the logic of such inferences began only a few decades ago. My own proposals in these regards are roughly outlined in earlier chapters. A more sophisticated and detailed approach is to be found in the work previously cited, to which corrections and emendations — and simplifications, I hope — are to be found in "The Reference Class."[2]

The other ingredient consists in the reasoned replacement of one way of talking by another. At an elementary, prescientific level, an example of this is the development of linguistic apparatus that allows for the articulation of the difference between *appearance* and *reality*. At the other end of the spectrum we have the introduction of specialized language for talking about the quaint objects and properties important to particle physics.

I say "reasoned" replacement: I have argued that such replacements are, or should be, made on rational grounds. In earlier chapters (and almost only there) I have offered my own crude proposals in this regard. One might refer to the kind of change of language that I envisage as "conceptual change"; this seems a rather grandiose term for what I take to be a fairly simple syntactical matter. Whether or not the proposals considered there can be made explicit and persuasive, I suggest that what we need here is only a way of choosing between explicit syntactical entities.

Presuppositions

My belief that both ingredients of the development of scientific knowledge can be represented as rational and articulate processes may be construed as an intuition, an article of faith, a speculation. We have

---

[1] H. E. Kyburg, *The Logical Foundations of Statistical Inference*, Reidel, Dordrecht, 1974.
[2] H. E. Kyburg, "The Reference Class," *Philosophy of Science*, 50, 1983, 374–397.

neither discovered a completely general and satisfactory logic of statistical inference nor generally acknowledged explicit criteria for the replacement of one way of talking by another. This has led a number of writers to conclude that the cogency of scientific argument rests on "presuppositions" or "assumptions." If the scientific method itself rests on "ultimate" presuppositions or assumptions, then those presuppositions themselves cannot be defended by scientific argument: they lie beyond the limits of scientific competence. If the very foundations of science lie beyond the limits of scientific competence, there follow consequences that may or may not be welcome.

It is possible to claim, for example, that science and religion can come into no conflict, since they represent alternative points of view resting on alternative presuppositions and assumptions. It is also possible to dismiss scientific knowledge that you would rather not accept by just saying that in that particular regard you decline to accept the presuppositions required. (And it seems that you may even excuse yourself from citing the presuppositions at issue!)

One of the most commonly alleged presuppositions of science concerns causality. This will serve to illustrate the claims made above. It is said that some assumption regarding causality is required for any scientific inference to be sound, and that (therefore) scientific inquiry is powerless to establish the principle of causality. Therefore faith in miracles is on a par with, though it depends on different presuppositions from, Newtonian mechanics. In conformity with the discussion of Chapter 11, I contend (a) that no presupposed causal principle is needed to establish the soundness of scientific inference, (b) that no causal principle that has been offered is sufficient to contribute to the soundness of scientific inference, (c) that no principle of causality in a general form is implied by our scientific knowledge, but (d) that if there is such a principle, it can be discovered—rendered acceptable—through empirical scientific inquiry, and finally, (e) that Newtonian mechanics is (within its approximations) warranted by experience in a way that miraculous violations of mechanics are not.

With respect to (a): statistical inference is one standard form of scientific argument. Some writers[3] have argued that some form of causal principle is required for the soundness of statistical inference, but their arguments have been unpersuasive. Given what many take to

[3]For example, Arthur W. Burks, *Chance, Cause, Reason*, University of Chicago Press, Chicago, 1977.

be the sketchiness of our present understanding of the logical structure of statistical inference, it is hardly surprising that these arguments are unpersuasive to many. On the other hand, there are many areas of science in which the major form of scientific knowledge is knowledge of statistical regularity. Whether or not there are assumed to be underlying causal mechanisms, these regularities themselves represent scientific knowledge about the world.

With respect to (b): the claim that there is a causal regularity underlying every event, if stated in a barefaced generality, is so loose as to be powerless to aid in the reconstruction of particular inferences. Without some constraints on the sorts of things that can enter into these regularities (and on the sorts of things that can be assumed to be regulated), we can get nowhere.[4] But as soon as the principle is stated in a more concrete and helpful way, for a particular case (e.g., there must be a bacterial or viral organism causing the collection of symptoms we call disease D), it becomes (i) grounded on scientific background knowledge and (ii) quite possibly false (D may be psychological or genetic or a cluster of diseases or some combination of the foregoing).

With respect to (c): it seems to me that we have far more evidence for tychism than for the causal principle—things never turn out exactly as we expect. We can always make excuses. But with all our scientific knowledge, we can predict, even with rough and ready precision, only in special and usually artificial circumstances. The faith that if only we had more scientific knowledge, or could base our predictions on more or more precise data, we could in principle predict the course of events (at least the course of nonhuman events) precisely strikes me as just as childish and ill-founded as the somewhat older faith that the course of human destiny is fatalistically determined by the whims of divine beings. It is a useless superstition.

With respect to (d): however, it cannot be denied that science and technology have given us the power to intervene in the course of events, to make them conform more nearly to our desires. We have uncovered many *specific* causal connections: bacterium B causes disease D; drug K causes the demise of bacterium B. Leaving to one side the essentially indeterministic character of quantum mechanics, we can *imagine* that there are no bounds to our ability to discover causal connections. Thus it might be the case that in the limit of scientific inquiry, some form of

---

[4] Assuming that the universe doesn't repeat itself, every universal event—state of the universe—has a unique predecessor and a unique successor.

causal principle might seem to hold true in the world. If that is so, it is something we can discover only by pursuing scientific inquiry toward its (temporal) limit.

With respect to (e): to accept Newtonian mechanics (in its place and degrees of approximation), as I have construed acceptance above, is exactly to rule out violations of its principles as impossible. It is not that miracles are to be construed as improbable, but that they are flatly impossible. Just as in the case of the theory of the measurement of length: an observational violation of the axioms is to be construed as evidence of error. Too much error, and we should change our language. But that hasn't happened yet.

The demand for "postulates" on which to found science reflects a failure of nerve. To be sure, we have yet to produce a totally adequate and widely accepted analysis either of statistical inference or of the replacement of one theory or law by another. To be sure, it is a speculative hypothesis, or an article of faith, that we will be able to do so. But if we don't *try* to produce a presupposition- and assumption-free analysis of scientific argument, we surely will not find one.

Realism and Intrumentalism

There is a lot of discussion in the philosophy of science on the issue of realism versus instrumentalism. Are the entities and processes referred to in our fancier scientific theories real—are they out there, doing at least approximately what our scientific theories say they do?—or is the whole edifice of science an enormous black box that allows us to crank out observational results from observational inputs, and no more? Do we not have to settle this question, on philosophical grounds, before we can even begin to understand science? I suggest that this is putting the cart before the horse. In specific cases, in ordinary discussion, we turn to science for answers to our ontological questions.

Are there really electrons? An affirmative answer is mandated by the body of scientific knowledge and evidence we now have. Given the evidence we now have, we have every reason in the world to believe that there are electrons. That is to say, we have every reason to speak a scientific language in which the term "electron" is taken to denote a relatively ordinary kind of object with relatively ordinary properties: an electron has a charge, a rest mass, a spin, and so on. To say this does not, of course, preclude the possibility that an expanded body of scientific knowledge and evidence might warrant a negative answer. But that is true of any scientific "fact."

An important distinction is made by Ian Hacking[5] between realism with regard to scientific objects (do electrons exist?) and realism with respect to scientific theory (do these objects have the properties that our theories say they do?). It seems much more reasonable to suppose that 50 years from now we will still be dealing with electrons than to suppose that we will have exactly the same physical view of their properties. Hacking ties this difference to experimental manipulation: we *use* electrons to investigate other physical objects. And, of course, this ties in with the view of causality propounded earlier: A causes B if you can use A to manipulate B.

Realism with respect to the properties that our "theoretical entities" have seems, like realism with regard to the entities themselves, to be a question that is appropriately answered by the science in question. And, also like the other question, to admit of degrees. There are some properties that we reasonably expect electrons will almost certainly be thought to have 50 years from now, and others that might have to be modified in the light of imaginable future experiments.

The answer to the general questions of whether we should suppose that some theoretical entities are real, and whether we should believe that they really have some of the properties we are currently attributing to them, is quite clearly "Sure." But that answer is not very enlightening. To get more interesting answers, we must ask case by case and even property by property.

To claim that the big question of instrumentalism versus realism can only be asked of one particular case at a time, and that it is scientific inquiry itself that must answer it, is not to say that the answer in any particular case is easy or obvious. The relevant scientific knowledge may be incomplete. There may be some reason to suppose that a theory involving X's is correct, but not a strong enough reason to preclude the development of an alternative theory involving entities that act *as if* they were X's.

This is exactly the region in which speculation reigns. A *good story* may give us "some reason" to believe in X's; another *good story* may give us "some reason" to believe that X's are fictional. If there is no rationally compelling evidence now, we must live with suspended judg-

[5]In Ian Hacking, *Representing and Intervening*, Cambridge University Press, Cambridge, 1983.

ment until there is. But at some point we may read in *Science* that a $48 million experiment at CERN has confirmed the existence of X's.

It is not only the lack of scientific knowledge that many render the answer to the question "Real or instrumental?" difficult. Reason and analysis may be required to interpret the deliverances of science. We need not go so far as to say that all ontological questions are internal to the special sciences. We need not go so far as to say that philosophy and analysis cannot contribute tentative answers to some of these questions. No sharp line can be drawn between science and philosophy. But in most cases our answers must be tentative, and the evidence on which these answers should be based is precisely *scientific* evidence.

Values

How about values and duties? Are they, too, to fall in the domain of natural science? What of Hume's "ought" and "is?" Can we derive "ought" from "is" after all? I won't claim that biology or psychology can tell us what we ought to do or what we ought to value. But at the same time, it seems clear that psychology and biology can tell us things that are *relevant* to our choices and actions.

In the first place, science can inform us of the consequences of our choices and actions. At least in some cases and in some degree, the consequences of our actions are relevant to their desirability and obligatoriness.

Rather more to the point, however, biology and psychology can give us insight into what we value and desire. It is naive indeed to say, "I know what I want and there's an end on't." On the other hand, it would be equally naive to suppose that scientific inquiry can resolve all questions of value, even leaving to one side questions of duty and obligation. My claim is not that. My claim is rather that we cannot *specify* the limits of science in this regard. We cannot say, "So far, and no farther." Knowledge of the kind of animal we are, knowledge about our reactions and satisfactions, can help to close the gap between what we think we want and what we *really* want.

But can scientific inquiry tell us what our duties and obligations are? That's what would confound Hume's observations.[6] Of course,

---

[6]David Hume, *A Treatise of Human Nature*, Oxford University Press, Oxford, 1960, Book III.

the pragmatic imperative—if you want to accomplish A, then you ought to do B presents no problems; indeed, it is one of the main functions of scientific knowledge to provide us with the grounds for accepting such imperatives.

Categorical imperatives seem to be quite different. Even here, however, if you accept the dictum that "ought" implies "can," scientific knowledge appears relevant. If, at the temporal limit of scientific inquiry, one approached a nomological net that totally constrained human action—that is, that eliminated human action in favor of predictable human behavior—there would be no room for obligation or duty except as (possibly) useful terms in that net.[7]

But this is a cheap shot. In the first place, we are nowhere near that limit, and I just finished arguing that it is an unwarranted and useless assumption to suppose that science is approaching it. In the second place, even supposing that LaPlace's demon lies at the temporal limit of scientific inquiry, he doesn't constrain us now: the nomological net that we know of leaves plenty of room for the notions of duty and obligation and freedom. In the third place, we were asking if scientific inquiry could contribute to our deontic knowledge; to say that in some hypothetical limit there would be no duties or obligations is no answer to that question. And in the fourth place, the dictum that "ought" implies "can," and its contrapositive "can't," implies "needn't" are questionable.

Nevertheless, even if we weaken the dictum to something like "ought to X" implies "by trying, you could come closer to doing X than you could otherwise," what we know of biology and psychology is still relevant to what we can take as obligatory. As in the case of value, it is not clear how far science can go in putting constraints on what we can reasonably take to be our duty; which is to say that it *is* clear that we *cannot* specify limits to the relevance of scientific knowledge to our conceptions of duty and obligation.

Science and Religion

The final region that is often alleged to lie beyond the limits of scientific knowledge is the religious. It is often said that religious questions are

---

[7]Much has, of course, been written about this question in ethics. There are a number of arguments purporting to show that a complete determinism regarding human behavior is compatible with action and responsibility. We need not enter into these questions here.

beyond the competence of science to answer; it is perhaps most often in connection with religion that allegations concerning the limits of science are invoked. There are a number of issues involved here, largely because the claims of religion can cover such a wide variety of areas.

At one extreme, there are head-on collisions between religion and science, as, for example, the conflict between creationism and evolutionary theory.

It has been argued that although there is a concrete factual issue involved, it cannot be resolved by conventional scientific procedures. The alleged reason is that conventional science has one set of "presuppositions and commitments," while creationism has a different set. Since these presuppositions and commitments lie beyond the limits of science, there is no non-question-begging way of adjudicating the claims of creationism and evolution. Thus, it is claimed, A may reasonably believe in creationism, and B, with the same evidence, may reasonably believe in evolution. One is reminded of the arguments (or claims) put forward by Kuhn and Feyerabend for the incommensurability of different scientific paradigms.

It has already been argued that this won't wash. The reasonableness of accepting scientific conclusions does not depend on presuppositions or assumptions. Indeed, one of the glories of science, and one of the ingredients of its progress, is precisely its propensity to question, to put to the test, and to repudiate unfounded assumptions. If natural science requires no presuppositions, then clearly one does not arrive at a different body of knowledge by making different assumptions; one arrives at superstition, rather than science, by adopting presuppositions that are unwarranted.

There are relatively few who would accept the pronouncements of religion (such as the one just mentioned) in areas that quite clearly fall in the domain of natural science. One does not consult the Bible for a theory of electron spin. There are many more who would accept the results of scientific inquiry, as far as it goes, but who would suppose that there are limits beyond which it cannot go. Beyond these limits, there are matters both of fact and of value where religious knowledge supervenes. The supreme matter of fact, of course, concerns the existence of God. The major matters of value concern sin and salvation.

The factual hypothesis is both vague and ambiguous. That doesn't make it true. The idea of divinity is rather amorphous, and every sect has its own god. Nevertheless, it is a factual hypothesis in the sense that, taken together with other things we know, it yields consequences in our bodies of practical certainties. Otherwise the whole claim is empty. But then we can go beyond Voltaire's cagey claim, "I have no need of that hypothesis."

Negative existential claims are not beyond the purview of science. (Consider perpetual motion machines.) We have outlived the ether, phlogiston, and a number of older chemical virtues. This is not to say that the theories embodying the assumed existence of those things have been refuted; I think the historical and philosophical analyses that suggest that out-and-out refutation does not occur in science are correct. But as the evidence mounts, there comes a point where it is no longer rational to attempt to operate within the framework of a theory (language) of a certain sort.

The evidence can become overwhelming that a given framework is ill-advised. It seems that the factual hypothesis in question falls into this category. Since it is vague, perhaps we can only say that it vaguely conflicts with those hypotheses (those linguistic frameworks) for which we have good evidence. But conflict is conflict, and it seems as rational to maintain a belief in phlogiston, appropriately diluted so as not to come into conflict with the facts of modern chemistry, as to maintain a belief in the existence of a dilute god who is careful not to intervene in the natural world revealed by science.

One can say much the same of sin and salvation, the existence of heaven and hell, judgment and damnation. Not only have we no use for these hypotheses, but vague and ambiguous as they are, they seem to conflict with what little many of us think we know about value and duty. One could still say that religion is a source and repository of important human values. That may be as it may be. Nevertheless, to the extent that scientific knowledge bears on these matters—and I have claimed that it does bear on them—the results of rational scientific inquiry must be given precedence over our initial prejudices and speculations, whether those stem from organized religious sources, superstition, or idiosyncrasy.

Conclusion

To sum up: I have argued—perhaps I should say "claimed"—that we can put no limits to the scope of scientific knowledge. In particular, I have argued that religion, ethics, and morals are impinged upon by scientific knowledge, and that we can set no limits to those impingements. With regard to science itself, I have claimed that science needs no assumptions or presuppositions, and thus that science has no foundations that are beyond the bounds of scientific inquiry. Logic itself may be construed as the outcome of empirical inquiry in the sense that its useful constraints reflect the interaction of nature and human nature. Logic, too, is our creation, and may be altered.

This is not to say that scientific knowledge embodies all answers to all questions, now or ever. It is to say that I can envisage no question to which scientific knowledge is in principle irrelevant.

This is quite different from saying (as Dewey seems to have suggested[8]) that the correct response to any question is to embark on a scientific inquiry. If I am faced with a moral dilemma, I must resolve it as best I can on the basis of what I know about both morals and the relevant facts now, rather than on the basis of what someone *might* know about morals someday, given unlimited scientific knowledge. It might be that that knowledge would be relevant if I had it; but I do not have it, and I must do the best I can with what I have.

By the same token, having granted factuality to the hypothesis of God's existence, I must admit that natural science might someday employ as its best theory one in which this hypothesis is an ingredient. But I have my doubts. I also doubt tht phlogiston will be revived. Predictions as to the course of future science aside, I attempt to believe what the empirical evidence rationally warrants. I take this to be the essence of scientific rationality, and I take the beliefs warranted at a certain time by scientific investigation to be the scientific *knowledge* of that time. Reason is the driving force behind science. Science is the embodiment of reason. And though there are competitors to scientific knowledge, and although that is a good thing (wild ideas enrich science as well as the rest of our lives), there are no *reasonable* or *rational* competitors.

---

[8] John Dewey, *Essays in Experimental Logic*, Dover, New York, 1953.

Name Index

Allais, Maurice, 237, 238
Archimedes, 4
Aristotle, 4, 13, 182
Ayer, A. J., 7, 17

Bayes, Thomas, 57, 116, 224, 225, 227, 228, 230, 238, 247, 242
Bergson, Henri, 5
Berkeley, George, 4, 15
Braithwaite, Richard B., 242
Bridgman, Percy, 9
Burks, Arthur W., 269

Carnap, Rudolf, 6, 38, 39, 50, 61, 62, 63, 92, 96, 259
Cartwright, Nancy, 172, 176
Cooley, John, 72
Copernicus, 4, 149
Craig, William, 113, 156

Dalton, John, 263
de Finetti, Bruno, 40
Democritus, 4, 263
Descartes, René, 4
Dewey, John, 276
Duhem, Pierre, 112, 113, 115, 148, 166, 167

Eddington, Arthur, 162
Einstein, Albert, 6, 8, 17, 35, 73, 111, 160, 162, 163, 164, 188

Ellsberg, Daniel, 237
Euclid, 4, 17, 22, 33, 34, 35, 111, 142

Feigl, Herbert, 6
Feyerabend, Paul, 159, 257
Fisher, Ronald A., 40, 205
Frege, Gottlieb, 6

Galileo, 4, 149, 150
Gardinfors, Peter, 190
Gibbard, Alan, 228, 229, 237
Giere, Ronald, 198, 199
Goldberg, Rube, 198
Gödel, Kurt, 6
Grünbaum, Adolf, 155

Haack, Susan, 34
Hacking, Ian, 37, 271
Hanson, Norwood Russell, 256
Harper, William, 228, 229, 237
Hegel, Georg, 6
Hempel, Carl G., 6, 38, 60, 197, 200, 201, 202
Hesse, Mary, 88
Hintikka, Jaakko, 38, 50
Hooke, Robert, 158
Hume, David, 4, 5, 15, 182, 185, 214, 272

Jaynes, E. T., 62
Jeffrey, Richard C., 63, 64

Kalish, Donald, 19
Kant, Immanuel, 5, 74, 182, 184
Kemeny, John, 38
Kepler, Johannes, 258, 264
Kripke, Saul, 213
Kronecker, L., 12
Kuhn, Thomas, 8, 10, 165, 166, 257

LaPlace, Pierre, 183, 188, 192, 195
Leeuwenhoek, Antonie van, 149, 151
Levi, Isaac Levi, 61, 63, 148, 204, 207, 239, 243, 250, 251
Lewis, David, 213, 217
Lindenbaum, A., 51
Locke, John, 4, 15

Mackinson, David, 190
Mill, John S., 17, 182
Montague, Richard, 19
Morgenbesser, Sidney, 243, 250, 251

Nagel, Ernest, 17
Newton, Sir Isaac, 6, 9, 73, 159, 160, 161, 162, 163, 164, 167, 174, 175, 177, 178, 179, 188
Nietzsche, Friederich, 6

Papineau, David, 202, 205, 209
Parmenides, 4
Pascal, Blaise, 37
Peano, Giuseppe, 30
Peirce, Charles S., 226
Plato, 13
Poincaré, Henri, 111, 113, 115
Poisson, Siméon D., 55

Popper, Karl R., 38, 39, 63, 78, 131, 132, 166, 259
Ptolemy, 149
Putnam, Hillary, 17

Quine, W. V. O., 23, 25, 31, 112, 113, 114, 115, 158, 166, 167

Reichenbach, Hans, 6, 37
Russell, Bertrand, 6, 183, 184, 212
Ryle, Gilbert, 72

Salmon, Wesley, 63, 197, 202
Schlick, Moritz, 6
Scriven, Michael, 204
Shortliffe, E. H., 38
Snell, W., 175
Snow, C. P., 11
Stalnaker, Robert, 213
Suppes, Patrick, 197, 199, 204

Tarski, Alfred, 21, 51
Tchebycheff, P. L., 54, 55
Toulmin, Stephen, 72

Ullian, Joseph, 158

van Fraassen, Bas, 150
von Mises, Richard, 6, 37

Wagner, Richard, 6
Whitehead, Alfred North, 6, 212
Wittgenstein, Ludwig, 7

Zadeh, Lotfi, 22

Subject Index

A priori constraints, 120
Acceptance, 61, 241, 264
Accuracy, 106, 109, 172
Acting as if, 243
Actual frequencies, 220
Addition of laws, 146
Additivity, 103, 105, 137–39
Aesthetic qualities, 262–64
Agency, 185, 187
Analyticity, 74
Approximation, 168, 171, 173, 178, 195
Arguments, 20
Arithmetic, 32, 33
Artificial intelligence, 10
Assumptions, 52, 268
Axiom schemata, 20
Axioms, 20, 74, 75, 259

Bayes' theorem, 228
Bayesian analysis, 71
Bayesian decision theory, 224, 225
Bayesian dominance, 46
Bodies of knowledge, 65, 66

Categorical imperative, 273
Causal connections, 218
Causal necessity, 212
Causality, 182, 186, 188, 195, 209
Chance, 39

Choice behavior, 181
Cognitive science, 10
Comparative ideality, 169
Computational simplicity, 156, 157
Conditional probability, 50
Conditionals, 190
Confidence interval, 154
Confidence methods, 233
Conforming probability functions, 249, 250
Context of discovery, 256, 259
Context of justification, 256, 259
Convention, 140, 171
Conventional choice principle, 130
Conventionalism, 113
Correlation, 204
Counterfactual conditionals, 189, 213, 217
Counterfactual probability, 222
Creationism, 265, 274
Crucial experiment, 162

Decision theory, 37, 179, 197
Deletion of laws, 146, 147
Derivability, 214
Determinism, 184
Deterministic programs, 208
Direct measurement, 106
Disagreement, 45
Disciplinary matrix, 10

Disposition, 199, 218, 219, 220, 222
Distribution principle, 91, 121, 138
Domination, 45, 180, 232

Economic man, 179
Empirical disciplines, 16
Empiricism, 74, 266
Empiricist, 4, 15
Epistemic homogeneity, 202, 207
Epistemic preference, 130
Epistemology, 250
Erroneous observation, 90, 104, 114, 119
Error, 76, 84, 102, 103–4, 118, 124–25, 138, 141, 189
Error distribution, 140
Error rates, 86, 92
Errors of measurement, 104, 135, 192, 193, 194
Euclidean geometry, 33, 35, 111
Evidential certainty, 68, 117, 141
Evidential corpus, 67, 116, 117, 154, 159, 214, 215, 216, 248
Evidential probabilities, 247
Evolution, 265, 274
Expectation, 228, 253
Expected utility, 180, 224, 225, 229, 232
Experience, 15, 74, 266
Experiment, 164
Explanation, 200, 201, 265

Facts and values, 14
First order logic, 18, 19
Formal disciplines, 16
Free choice, 230
Full (r/s) belief, 246
Full belief, 241, 242, 250, 251
Fundamental idealization, 177
Fuzzy logic, 22

Generalization, 115
Geometry, 21, 33, 34, 111

History of science, 257
Homogeneity, 202

Icosahedron, 220
Ideal agents, 235
Ideal boundary condition, 175
Ideal coin, 221, 223
Ideal gas, 172, 222, 223
Ideal gas law, 171–74, 177, 221
Ideal laws, 142
Ideal stochastic object, 222, 223
Idealization, 168, 175
Incorrigible observations, 134
Independence, 194
Indirect measurement, 106
Induction, 53, 58, 266
Inductive considerations, 115
Inductive explanation, 201
Inductive generalization, 58, 116
Inductive logic, 259
Inference structures, 44
Instantial induction, 59, 68, 72
Instrumentalism, 150, 157, 265, 270, 271
Interpersonal uniformity, 82
Interval-valued probability, 231

Judgment, 23, 74, 77–79, 103, 266
Justification, 76, 259

Languages, choice between, 126, 229, 255, 260
Law of linear thermal expansion, 142–44
Laws and theories, 115
Level of acceptance, 64, 241
Linguistic constraints, 257
Linguistic convention, 144
Logic, 17
Logic of discovery, 259
Logical knowledge, 75
Lottery paradox, 64

Magnitude, 96, 97, 99, 101–3
Manipulation, 188, 210
Margins of error, 164
Mass, 159, 160, 161
Mass behavior, 37
Mathematics, 4, 11, 17
Maximal specificity, 200, 201, 202, 205, 207
Maximax strategy, 233
Maximin strategy, 233
Maximizing expectation, 227
Maximizing expected utility, 179, 180, 181, 225, 231, 237, 242
Measure, 37
Measurement, 98, 172, 189, 190, 192
Metalanguage, 18
Minimization principle, 90, 120, 138
Modal logic, 212, 213
Model, 21, 32, 55
Modus ponens, 20

Necessity, 212, 214
Newcomb's problem, 209
Nomic generalizations, 72
Nomic induction, 59
Nomic necessity, 119
Nomological explanation, 201
Nondemonstrative arguments, 36
Nondemonstrative inference, 58
Nonmonotonic inference, 40, 214
Normal distribution, 136
Normative idealizations, 179

Object language, 18
Observation, 76, 79, 81, 82
Observation and experiment, 15
Observation reports, 88, 89, 90, 128, 164, 263
Observation statements, 68, 77, 88, 89, 113, 128, 164
Observation vocabulary, 114, 149, 156
Observational error, 85

Ontology, 96, 265, 270
Operational definition, 107
Operationalism, 9
Ordinary language, 79

Paradigm, 10, 274
Paradoxes, 237
Partial belief, 241, 242
Particular sentences, 76
Phrenology, 83, 149
Plausible probabilities, 227
Positivism, 7
Possibility, 211, 212
Possible worlds, 213, 214, 217
Powers, 3, 184, 199, 210, 218
Practical certainty, 66, 68, 118, 152, 159, 176, 215, 226, 235
Practical corpus, 116, 227, 231, 241, 248
Prediction, 173, 189
Predictive information, 140
Predictive observational content, 130, 142, 152–54, 156, 176, 181, 208, 258, 288
Preferences, 235
Presuppositions, 52, 265, 267, 268
Principle B1, 245
Principle B2, 247
Principle B3, 249
Prior probabilities, 224, 228
Probabilistic causality, 198, 199, 209
Probabilistic independence, 51
Probabilistic inference, 53
Probability, 180, 197, 214, 215, 225, 231, 240, 241, 252–54
Probability$_1$, 39
Probability$_2$, 39
Progress, 7, 8
Propensity, 39

Quantities, 94, 135
Quantum mechanics, 6, 19, 152, 183, 195, 196, 210, 222

Random distribution, 198
Random error, 109
Random member, 44, 162, 223, 225, 231
Random quantities, 96, 135
Random variable, 95, 96, 135
Randomness, 44, 50, 123
Rare events, 253
Ratio of risk to reward, 245
Rational action, 254
Rational agent, 179
Rational and justified, 5
Rational belief, 250, 254, 255
Rational corpus, 43, 48, 76, 263
Rational full (r/s) belief, 252
Rationality, 4, 180, 241, 248, 254, 256
Real possibility, 254
Realism, 150, 265, 270, 271
Reason, 276
Reference class, 202, 207, 220, 229, 230
Reference term, 203, 207
Refutation, 131, 166
Relative frequency, 37
Relativity, 152, 157
Relevance, 52
Reliability, 77, 85, 158
Religion, 268, 273, 274, 275
Replacement, 190
Replacement of a theory, 148
Replacement of laws, 146
Representativeness, 43
Responsibility, 186, 187
Revolution, 165

Sample inclusion, 47
Scientific inference, 268
Scientific judgment, 256
Scientific method, 3, 268
Scientific orthodoxy, 262
Scientific rationality, 276
Screening off, 204, 205

Security, 76
Semantic content, 150
Semantics, 18
Sense data, 76
Sets of distributions, 55, 56
Simplicity, 111, 112
Simpson's paradox, 205
Sociology of science, 257
Special relativity, 158
Stakes, 244, 245, 251
Statistical causation, 195
Statistical explanation, 200, 201
Statistical generalizations, 76
Statistical induction, 58, 59, 69
Statistical inference, 57, 255, 266, 268, 269
Statistical laws, 196, 197
Statistics, 52
Stochastic independence, 51
Stochastic knowledge, 252
Subjective probability, 38, 62
Subset, 45
Superstition, 187, 209
Systematic error, 108

Testimony, 78
Theoremhood, 75
Theoretical induction, 59
Theoretical terms, 259
Trivial semantic conventionalism, 155
Truth in a model, 21
Tychism, 269

Uncertainty, 36
Underdetermination, 112, 166
Understanding, 265
Uniformity, 184
Unit, 99, 102, 135, 155
Universal causal determinism, 183
Universal causation, 182, 184, 219
Universal deterministic causation, 195

Universal generality, 119
Universal generalization, 72, 76, 166
Universal induction, 59, 71
Upper and lower probabilities, 233
Ur-corpus, 117
Utility, 180, 181, 234, 241
Utility theory, 235

Vagueness, 22
Value theory, 236
Values, 265, 272
Veridical observations, 90, 119

Weak deductive closure, 48, 67
Web of belief, 112